Building Pathology and Rehabilitation

Volume 28

Series Editors

This book series addresses the areas of building pathologies and rehabilitation of the constructed heritage, strategies, diagnostic and design methodologies, the appropriately of existing regulations for rehabilitation, energy efficiency, adaptive rehabilitation, rehabilitation technologies and analysis of case studies. The topics of Building Pathology and Rehabilitation include but are not limited to - hygrothermal behaviour - structural pathologies (e.g. stone, wood, mortar, concrete, etc…) - diagnostic techniques - costs of pathology - responsibilities, guarantees and insurance - analysis of case studies - construction code - rehabilitation technologies - architecture and rehabilitation project - materials and their suitability - building performance simulation and energy efficiency - durability and service life.

Fernando A. N. Silva · Rodrigo F. Roma ·
Mohamed K. Bourbatache · Mahfoud Tahlaiti ·
João M. P. Q. Delgado · António C. Azevedo

Advanced Chemical and Creep Modeling for Alkali-Aggregate Reaction in Concrete

Springer

Fernando A. N. Silva
Department of Civil Engineering
Catholic University of Pernambuco
Recife, Pernambuco, Brazil

Mohamed K. Bourbatache
LGCGM
INSA Rennes
Rennes, France

João M. P. Q. Delgado
CONSTRUCT-LFC, Faculty of Engineering
University of Porto
Porto, Portugal

Rodrigo F. Roma
Department of Civil Engineering
Catholic University of Pernambuco
Recife, Pernambuco, Brazil

Mahfoud Tahlaiti
GeM—Centrale Nantes
Nantes, France

António C. Azevedo
Departamento de Construction
Instituto Federal de Pernambuco (IFPE)
Recife, Pernambuco, Brazil

ISSN 2194-9832 ISSN 2194-9840 (electronic)
Building Pathology and Rehabilitation
ISBN 978-3-031-53979-4 ISBN 978-3-031-53980-0 (eBook)
https://doi.org/10.1007/978-3-031-53980-0

This Springer imprint is published by the registered company Springer Nature Switzerland AG
The registered company address is: Gewerbestrasse 11, 6330 Cham, Switzerland

Paper in this product is recyclable.

Preface

Concrete structures suffering from Alkali Aggregate Reaction (AAR) are also affected by other deleterious mechanisms such as creep and shrinkage. In Brazil and many other countries worldwide, numerous cases of building foundations and concrete dams were investigated due to the damage associated with AAR. Macroscopic AAR models must have to consider the influence of the main environmental parameters, such as temperature, saturation degree and the speed of the evolution of the chemical reaction. To be relevant in structural applications, concrete creep models have to consider several important phenomena, such as nonlinearity, multi-axiality, and thermal and drying effects.

In order to prevent those pathologies, plan rehabilitation works, or create new design procedures, the numerical simulation through finite elements method (FEM) is recognized as a very useful tool. This work aims to implement a chemical model to simulate the advancement of the reaction and a mechanical model to simulate creep, and shrinkage phenomena in COMSOL Multiphysics®, as a way to reassess concrete structures suffering from those mechanisms. Both models were implemented separately to evaluate their responses and compare them with the theoretical results and experimental benchmarks proposed by the developers of these models and currently in use on other commercial software like CAST3M and Code_Aster. The models were implemented in COMSOL separately in order to evaluate their responses, comparing them to theoretical results and experimental benchmarks proposed by the developers of such models. The chemical model to reproduce AAR was compared to the theoretical results shown by the developer of the model, and the creep model was compared to experimental data regarding creep strains in different directions of a specimen under various loads and different temperatures.

The results of the simulations were in reasonably good agreement with the results taken from the literature, with a maximum error of 10%, showing that the implementation in COMSOL was a success. Although the shrinkage model didn't show any good results, and some numerical instability was encountered when the coupling of such models was done, and to fully reproduce a concrete structure affected by those deleterious phenomena, a coupled chemo-mechanical-damage model in a rheological framework must be implemented, which will be the aim of the next research projects.

Recife, Brazil — Fernando A. N. Silva
Recife, Brazil — Rodrigo F. Roma
Rennes, France — Mohamed K. Bourbatache
Nantes, France — Mahfoud Tahlaiti
Porto, Portugal — João M. P. Q. Delgado
Recife, Brazil — António C. Azevedo

Contents

Chapter 1
Introduction

The concrete is probably the most used civil construction material in the modern world. Developing countries such as China, India and Brazil stand out as the largest consumers of this material in the execution of buildings and infrastructure. The durability of concrete structures must be guaranteed in order to prevent expensive maintenance or even failure of those structures (Martin et al. 2012b).

The behavior of concrete includes the viscosity, elasticity and plasticity, and it can be characterized as an elasto-viscoplastic material, which its strains can be due to shrinkage, creep and damage (Leonhardt and Mönning 1977). The strains linked to shrinkage and creep are known as the two main factors affecting the durability of the structures in the long term. To better understand how those strains occur over time, many experimental programs, formulations and models with the objective of predicting the concrete behavior were developed (Abdellatef et al. 2015).

In addition, expansion is another type of deformation affecting concrete. The chemo-physical expansion processes activated by certain environmental conditions can lead to significant loss in strength and stiffness of many concrete structures. When those environmental conditions are presented, damage due to expansion might appear, and those expansions are due to one or more pathologies affecting the concrete structures (Scrivener 2009; Comi and Perego 2011). The inadequate durability of the concrete is perceived by the pathological manifestations, whose agents can reside inside itself or come from the external environment. Internally, these agents can be found in the chemical and mineralogical composition of cement or aggregates. Externally, aggressive substances penetrate the interior of concrete structures through the capillary network (connected porous matrix) in the form of gases, vapors or liquids, leading to chemical deterioration, which may involve the formation of expansive products (e.g., the alkali-aggregate reaction).

Some pathologies such as Alkali-Aggregate Reaction (AAR) and Delayed Ettringite Formation (DEF), external sulfate attack and even frost, are considered as Internal Swelling Processes (ISP). The ISP cause expansion on the affected concrete, leading to cracking, decrease of its mechanical properties and loss of early durability (lifespan

F. A. N. Silva et al., *Advanced Chemical and Creep Modeling for Alkali-Aggregate Reaction in Concrete*, Building Pathology and Rehabilitation 28,
https://doi.org/10.1007/978-3-031-53980-0_1

of concrete structures). These pathologies often cause severe problems in terms of serviceability, structural integrity and sustainable operation of the concrete structures. The two reactions (AAR and DEF) are often observed acting simultaneously on concrete structures, implying the precise understanding of these processes acting in two scenarios: alone and combined (Martin et al. 2012a, b).

The AAR is a phenomenon involving chemical reactions in which some constituents of the aggregate react with the alkaline hydroxides dissolved in the solution of the concrete pores, and it leads to cracking and loss of durability in cement-based composites. AAR is a chemical reaction in either concrete or mortar between hydroxyl ions (OH^-) of the alkalis (sodium and potassium) from hydraulic cement (or others sources), and certain constituents of some aggregates; under certain conditions deleterious expansion of concrete or mortar may result (Rajabipour et al. 2015).

Basically, four conditions must be satisfied in order to generate AAR on concrete: the presence of reactive aggregate, an adequate level of alkalinity, sufficient moisture in the pores of concrete, and high temperature. The understanding of AAR is very complex and such processes have an important role on the pathologies found on big structures such as bridges, railroads, dams and building foundations. The damage caused by AAR goes from localized cracking to generalized failure and performance degradation. The effective resolution of this problem is also very complex since the description of the mechanism of the reaction is not yet endorsed by conclusive theoretical formulation accepted by worldwide scientific associations, and the experimental tests for careful observation of the phenomenon requires a considerable amount of time and resources.

As the AAR is a process that takes many years to produce visible damage, it is very important to predict the behavior of the AAR-affected concrete structures, done in a way that is possible to evaluate the stability and durability of such structures. In the other hand, the numerical simulations of AAR-affected concrete structures generally neglect the interactions between AAR, creep and shrinkage (Grimal et al. 2007; Comi and Perego 2011; Pan et al. 2012).

1.1 Relevance of the Topic

As time goes by, society needs larger and higher structures than before in order to continue the process of development of cities and their infrastructures. Such megastructures (e.g., large raft foundations, dams, long tunnels, and bridges) use a huge volume of concrete in its execution, and such amount of concrete needs a large quantity of aggregate, but sometimes non-reactive aggregates aren't available locally—near to the construction site—and shipping it wouldn't be feasible. In addition, reactive aggregates are readily and abundantly available at lower cost and reduced environmental impact when compared with non-reactive materials (Thomas et al. 2013). But the usage of reactive aggregates may lead to the appearance of AAR if all the environmental conditions are satisfied.

The deterioration caused by the ISP leads to high-cost maintenance or rehabilitation of the affected concrete structures, and many countries all over the world have been reporting evidences of DEF and AAR on these kinds of infrastructures and buildings. According to Taylor et al. (2001) and Hobbs (1999), in many of these cases the DEF happened simultaneously with the AAR, when siliceous aggregates were used on concrete. In the literature there are a lot of reports showing those two phenomena—DEF and AAR—often manifested together since both are activated by high temperatures on the curing stage, followed by exposure in a high humidity environment.

In many countries a large amount of resources has been applied to the repair and maintenance of concrete structures affected by the AAR. For example, according to Oberholster, quoted by Kuperman (2007), South Africa had spent the amount of U$ 350 million only between the 1970s and 1996. In the year of 2008, the organizers of the 13th International Conference on Alkali-Aggregate Reactions in Concrete (ICAAR) estimated the annual cost to rehabilitate and substitution of constructions affected by ISR is U$ 2.6 billion.

Another good example is the case of the Elgeseter Bridge, in Norway, built between 1949 and 1951 and costing kr$ 1.5 million at the time. In 1989 the bridge was already damaged due to shrinkage, creep and corrosion of the reinforcements concrete columns, but in the early 90s many other damage due to AAR was also verified, and the cost to repair all the bridge in 5 years was around kr$ 6.8 million at the time (Jensen 2004).

Such huge amounts of money are a big concern to the civil construction industry and also the governments all over the world, leading to the creation and funding of many research centers (e.g., LMDC-INSA) in order to develop technology suitable for characterization, mitigation and repairing of the affected structures. The main research centers are located in France, Canada, the United Kingdom, Germany, and the United States of America.

According to Martin et al. (2012b), there are three components to the investigation of structures suffering from ISP:

- Chemo-physical characterization focusing primarily on the materials (cement and aggregates);
- Computational modeling of the evolution of damage and assessing the structural response of the structure;
- Managing the structure.

The presenting research project, a partnership between UNICAP (Universidade Católica de Pernambuco—Recife, BR) and ICAM (Institut Cathólique de d'Arts et Métiers—Nantes, FR) is focused on the computational modeling, as the numerical simulation is an adequate instrument to quantify the extension of the damage caused by the AAR, assisting on the decision making regarding structural recovery works and also serving as a basis to modify some structural design criteria. In possession of all the information shown above, the research of numerical modeling of ISP affecting concrete structures must be done with caution and precision in order to mitigate the effects of these phenomena still on the design project stage, avoiding the high costs of maintenance and rehabilitation on those affected structures.

This research program was done during the COVID-2019 pandemic scenario, a very hard time for everyone worldwide. The ones involved in this work had put a lot of effort in order to maintain the schedule of the program, and most of the weekly reunions were done in a virtual platform.

1.2 Objectives

This work aims to perform a numerical simulation of the chemical and creep behavior of concrete affected by Alkali-Aggregate Reaction (AAR) on an isotropic non-linear bidimensional model using the software COMSOL Multiphysics®. This software is based on Finite Element Method (FEM) for the solution of mechanical analysis for linear and non-linear problems. The specific objectives are:

- Organize and present a literature review about AAR chemistry modelling, as well as creep and shrinkage modelling.
- Implement the chemical model proposed by Morenon (2017) in COMSOL Multiphysics®, which calculates the pressure of the gel regarding the volume of gel created and the advancement of the reaction, comparing the results in order to validate the implementation.
- Implement the creep and shrinkage models proposed by Sellier et al. (2016) in COMSOL Multiphysics®, which relates the permanent and reversible creep strains to the elastic strain in any direction and also to some fitting material parameters obtained in creep tests. The validation of the implementation was made by comparison with the experimental benchmark proposed by the developers of the model.

1.3 Methodology

For the development of this work, primarily it was done a literature review in order to search for the most advanced and suitable numerical models related to AAR, creep and shrinkage. The models developed and published by the LMDC-INSA Toulouse, coordinated by the professors Alain Sellier, Stéphane Multon and others, were selected due to its background: they were calibrated by extensive experimental programs and experimental data found in literature. The very up-to-date models showed results in very good agreement to the experimental results.

For the chemical model, it was selected the model proposed by Morenon (2017), which takes into account the main environmental and material parameters that influence on the advancement of the reaction and the ASR gel pressure generated. For the creep and shrinkage behavior, it was selected the model proposed by Sellier et al. (2016).

Those models were implemented on COMSOL Multiphysics®, a commercial cross-platform finite element analysis, solver and multiphysics simulation software. It allows conventional physics-based user interfaces and coupled systems of partial differential equations (PDEs). COMSOL provides an ideal and unified workflow for electrical, mechanical, fluid, acoustics, and chemical applications.

When the models were successfully implemented, the modules were validated by comparison with experimental and numerical benchmark found in literature. The coupling of all the models showed numerical instability, therefore it's the full coupled chemo-mechanical model is not presented in this work.

For the readers, probably there is still a question remaining: why, amongst all the FEM softwares available, the author chose COMSOL? The answer for this question remains on the procedures adopted to develop this research, as those chemo-mechanical models have already been implemented in other commercial softwares like CAST3M and Code_Aster. Also, COMSOL is able to simulate sensors in concrete structures, as seen on the works of Alcantara Jr. and Gonçalves Jr. (2015) and Pomarico et al. (2016), hence it would make possible—in the near future—to simulate a real concrete structure affected by ISP and the sensors inside of it, comparing the real data with the numerical simulations, providing an enormous contribution to the scientific community and civil construction industry.

In resume, in this work it is presented in the first chapter a general introduction about the problematic of the degradation of concrete structures, the importance of studying it and some costs related to those problems, which highlights the justification behind this research. Also, it's presented the objectives related to the research project and the methodology used in the development of this work. At least, it's shown the organization of the report.

The second chapter is about the literature review of this research, regarding the main concepts on the two subjects of the work: alkali-aggregate reaction (AAR) in concrete and the mechanical behavior induced by creep and shrinkage. The first part, which is about AAR, it's presented a historical background and the problem in Brazil, followed by a brief explanation about the formation of the gel produced by the reactions, a resume about the swelling and cracking mechanisms related to AAR, the factors which influence in this reaction and the main numerical models related to this problem. The second part, regarding the mechanical behavior of concrete, it's first presented the main concepts about creep and shrinkage, the factors which influence those phenomena and the main mathematical models, and secondly it's presented a brief concept about the poro-mechanical theory and the poro-mechanical models for swelling products in concrete structures.

The third chapter concerns about the materials and methods used in these research project, regarding the numerical models for both AAR (Morenon 2017), creep and shrinkage (Sellier et al. 2013), explaining the required formulation and the main material parameters and its suggested values. Next, it's presented the implementation of those models in the software COMSOL Multiphysics®, and a simple step-by-step procedure for any researcher to do the same.

The fourth chapter concerns about the results of such simulations, according to the theoretical behavior and some experimental benchmarks available on the works of the developers of the models presented in Chap. 3.

The fifth and last chapter concerns the main conclusions about the work, regarding the limitations and the positive points encountered by the author, and of course some perspectives for future works.

References

Abdellatef M, Alnaggar M, Boumakis G, Cusatis G, Di-Luzio G, Wendner R (2015) Lattice discrete particle modeling for coupled concrete creep and shrinkage using solidification microprestress theory. In: CONCREEP 10: mechanics and physics of creep, shrinkage, and durability of concrete and concrete structures, Vienna, pp 21–23

Alcantara Jr NP, Gonçalves Jr L (2015) Simulation of an ECT sensor to inspect the reinforcement of concrete structures. In: Proceedings of the 2015 COMSOL conference, Curitiba, p 6

Comi C, Perego U (2011) Anisotropic damage model for concrete affected by alkali-aggregate reaction. Int J Damage Mech 20(4):598–617

Grimal E, Sellier A, Le Pape Y, Bourdarot E (2007) Creep, shrinkage, and anisotropic damage in alkali-aggregate reaction swelling mechanism—part I: a constitutive model. ACI Mater J 227–235

Hobbs DW (1999) Expansion and cracking in concrete associated with delayed Ettringite formation. In: Ettringite: the sometimes host of destruction. American Concrete Institute, Michigan, pp 159–182

Jensen V (2004) Alkali-silica reaction damage to Elgeseter Bridge, Trondheim, Norway: a review of construction, research and repair up to 2003. Mater Charact 53(2–4):155–1770

Kuperman SC (2007) Considerações sobre fluência de concretos. Téchn Rev Tecnol Constr 125:58–63

Leonhardt F, Mönning E (1977) Construções de Concreto. Princípios Básicos do Dimensionamento de Estruturas de Concreto Armado, vol I, 2nd edn. Interciência Ltda, Rio de Janeiro

Martin RP, Toutlemonde F (2012a) Theoretical and experimental validation of a simple method to reproduce representative DEF-prone conditions in laboratory. Mater Struct 46(8):1245–1255.

Martin RP, Metalssi OO, Toutlemonde F (2012b) Modelling of concrete structures affected by internal swelling reactions: couplings between transfer properties, alkali leachning and expansion. In: 2nd international conference on microstructure related durability of cementitious, Amsterdam, p 8

Morenon P (2017) Modélisation des reactions de gonflement interne des bétons avec prise en compte des couplages poro-mécaniques et chimiques. Ph.D. thesis, Université Toulouse 3 Paul Sabatier, France, p 272

Pan JW, Feng YT, Wang JT, Sun QC, Zhang CH, Owen DRJ (2012) Modeling of alkali-silica reaction in concrete: a review. Struct Civ Eng 6:1–18

Pomarico AA, Roselli G, Caltabiano D (2016) Modelling of viscoelastic phenomena in concrete structures. In: Proceedings of the 2016 COMSOL conference, Munich

Rajabipour F, Giannini E, Ideker JH, Dunant DF (2015) Alkali-silica reaction: current understanding of the reaction mechanisms and the knowledge gaps. Cem Concr Res 76:130–146

Scrivener KL (2009) Importance of microstructural understanding for durable and sustainable concrete. In: Concrete repair, rehabilitation and retrofitting II, pp 13–20

Sellier A, Casaux-Ginestet G, Buffo-Lacarrière L, Bourbon X (2013) Orthotropic damage coupled with localized crack reclosure processing—part I: constitutive laws. Eng Fract Mech 971:148–167

Sellier A, Multon S, Buffo-Lacarrière L, Vidal T, Bourbon X, Guillaume C (2016) Concrete creep modelling for structural applications: non-linearity, multi-axiality, hydration, temperature and drying effects. Cem Concr Res 79:301–315

Taylor HFW, Famy C, Scrivener KL (2001) Delayed Ettringite formation. Cem Concr Res 31:683–693

Thomas MDA, Fournier B, Folliard KJ (2013) Alkali-aggregate reactivity (AAR) facts book. Report No. FHWA-HIF-13-019. U.S. Department of Transportation—Federal Highway Administration, Springfield, VA, p 211

Chapter 2
Literature Review

For well-designed concrete structures, the main source of damage is the shrinkage, creep and expansion processes, which can act combined or not. The expansion processes are often called Internal Swelling Processes (ISP), which are pathologies that degrade the concrete structures due to the formation of an intrapore product, and its swelling is responsible for damaging the solid skeleton of concrete.

This chapter is then divided in two parts, the first one is dedicated to the AAR and the second part is dedicated to the main mechanical behaviors of concrete which are necessary to numerically assess a concrete structure. Those mechanical behaviors can be resumed in creep and shrinkage, poro-mechanical theory and the main poro-mechanical models related to AAR.

In the subject of AAR, first is provided an historical background and an overview of the problem in Brazil, followed by the main reaction mechanisms and the most significant parameters on the formation of expansive product (AAR gel), and for the last is related the main mathematical models found in the literature related to the formation of the gel and its expansion.

In the subject of mechanical behavior, first is developed a brief concept of creep and shrinkage, followed by the theory of poro-elasticity and the models making it possible to numerically simulate the damage on concrete structures, and the effects of swelling stresses on the porous media of concrete.

2.1 Alkali-Aggregate Reaction (AAR)

In this topic, it's presented a brief historical and theoretical review of the concepts linked to AAR. The factors which influence the occurrence of those phenomena are commented on as well. This explanation is necessary in order to better understand the chemical model and its main material parameters presented in Chap. 3.

F. A. N. Silva et al., *Advanced Chemical and Creep Modeling for Alkali-Aggregate Reaction in Concrete*, Building Pathology and Rehabilitation 28,
https://doi.org/10.1007/978-3-031-53980-0_2

2.1.1 Historical Background and the Problem in Brazil

In the 1930s, a several numbers of pathological manifestations with expansive behavior were found on bridges and pavements in USA, and the real causes were not known at the time (Hobbs 1988). Since Stanton (1940) proved through experimental tests that aggregates derived from natural rocks aren't necessarily chemically inert materials (as they were supposed to be), the scientific community knows that the mineralogical constituents of the aggregates react with the alkalis from Portland cement. This chemical reaction was called Alkali-Aggregate Reaction (AAR).

At the beginning of 1980s, an important number of concrete structures had their damages associated to the expansive effect of Alkali-Silicate gel produced by AAR. At the time, the concern on the phenomenon's consequence motivated the beginning of more careful scientific investigations. The problem was put on the spotlight in a way that the volume and diversification of scientific research on this subject has been progressively expanded, leading to the growing discovery of new cases (Thomas et al. 2013).

The damage caused by AAR can be found without difficulties in concrete structures throughout the world, and today more than 50 countries have reported the occurrence of this phenomenon. In fact, some researchers believe that there are characteristic damages produced by AAR in concrete structures that are erroneously attributed to other causes, and it leads the scientific community to think that the seriousness of this problem (AAR) is even greater than they thought it was.

In Brazil, the firsts scientific works about the reactive potential of the aggregates are dated from 1960s when Heraldo Gitahy did extensive research on pozzolanic materials for the CESP's Urubupungá complex, which used calcined clay to mitigate AAR (Pacelli 1999). This technique was a success because until now there are no registries of this kind of reaction affecting the buildings. However, it was only in 1997 that the problematic of AAR gained more attention with the creation of the 1st Symposium on Alkali-Aggregate Reactivity in Concrete Structures. Silveira (1997) showed that in Brazil, the first records of concrete structures affected by AAR are dated from 1946 in the CEMIG's Peti Dam, and the aggregate used to fabricate the concrete was originally from gneiss rock.

Since 2004, the attention of the scientific community in Pernambuco (PE-BR) has turned to civil constructions in the Recife's Metropolitan Region (RMR) that was motivated by the collapse of the building Areia Branca. An extensive inspection was carried out on the foundation structures and at the end of several diagnoses made by many professionals, some signs of AAR were verified, although this pathological manifestation was not the cause of the collapse of the building. In Pernambuco there is a significant number of constructions proven to be affected by AAR. The cases registered are:

(i) sSx foundation slabs for refrigeration towers;
(ii) Tapacurá's Dam;
(iii) Paulo Guerra's bridge;

(iv) Dozens of foundation blocks (pile caps) from buildings situated on the RMR (Andrade 2006; Andrade et al. 2006a; Andriolo 2000; Brito Júnior and Ferro 2016).

The expansions of the aforementioned structures can be attributed to the usage of a large rock deposit located very close to a geological fault, and such deposit is a source of extraction of aggregates for concrete. It is known that quartz minerals in regions subjected to tectonic movements present their crystals with a deformed geometric arrangement, which is one of the varieties responsible for the slowest expansion rates that can last for many years (Madureira 2007).

Since the collapse of the building Areia Branca until 2006, more than 30 cases of AAR were found on the RMR. According to Andrade (2006), Andrade et al. (2006a, b) and Pecchio et al. (2006), the major causes are:

(i) the existence of aggregates containing extremely deformed, microcrystalline and recrystallized quartz;
(ii) low depth of the water table (near to the ground surface);
(iii) availability of soluble alkalis (specially potassium);
(iv) high ambient temperatures. Unlike the classic cases in the literature, where deleterious expansions occur over long periods of time, some constructions on the RMR showed deleterious cracks just in three years of construction.

In some cases, the expansion caused by the AAR is so severe that the internal tensile stresses in the concrete reached extremely high values (Pecchio et al. 2006), as can be seen by the rupture of the stirrup shown in Fig. 2.1. The magnitude of these expansions caused by the AAR in the constructions of RMR can also be visualized in the photographs shown in Fig. 2.2.

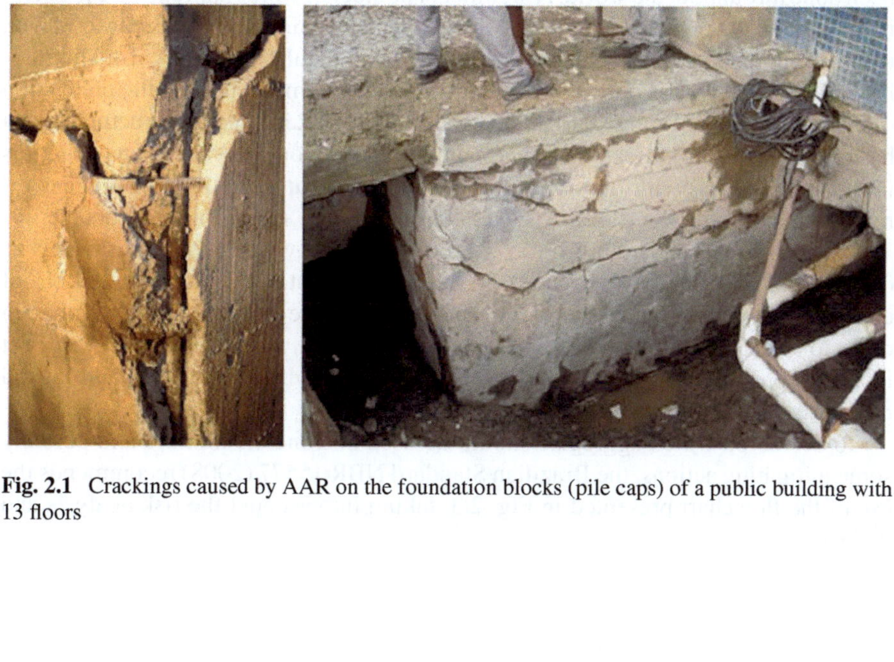

Fig. 2.1 Crackings caused by AAR on the foundation blocks (pile caps) of a public building with 13 floors

Fig. 2.2 Crackings caused by AAR on buildings at the RMR

After 80 years since Stanton's first works, many other researches programs had been done throughout the world. A significant number of works approach the problem in a very efficient and clear way on how to avoid the reaction before it is consolidated, using techniques which consists on the use of mineral and chemical addictions, limitation of the alkali content of cement, showing the importance on the prevention of the phenomenon. With regard to the treatment of AAR in concrete structures, what has been seen are empirical measures such as: (i) injection of cementitious materials in the cracks; (ii) monitoring of the expansions; (iii) cuts to relieve tension caused by the expansive products of AAR; (iv) restructuring of structural members etc.

The Brazilian Standard NBR 15577 (2008) shows how to perform a risk analysis on the possibility of the occurrence of AAR by taking into account the conditions of exposure, dimensions and the importance of the structure or the concrete member, in order to indicate a possible preventive action. However, the document does not yet include repairs and/or mitigating actions in situation where the AAR pathological manifestations are already happening.

To decide on which aggregate should be used on concrete mixture and possibly some mitigating actions, the Brazilian Standard NBR 15577 (2008) recommends the use of the flowchart presented in Fig. 2.3, taking into account the risk analysis and its factors.

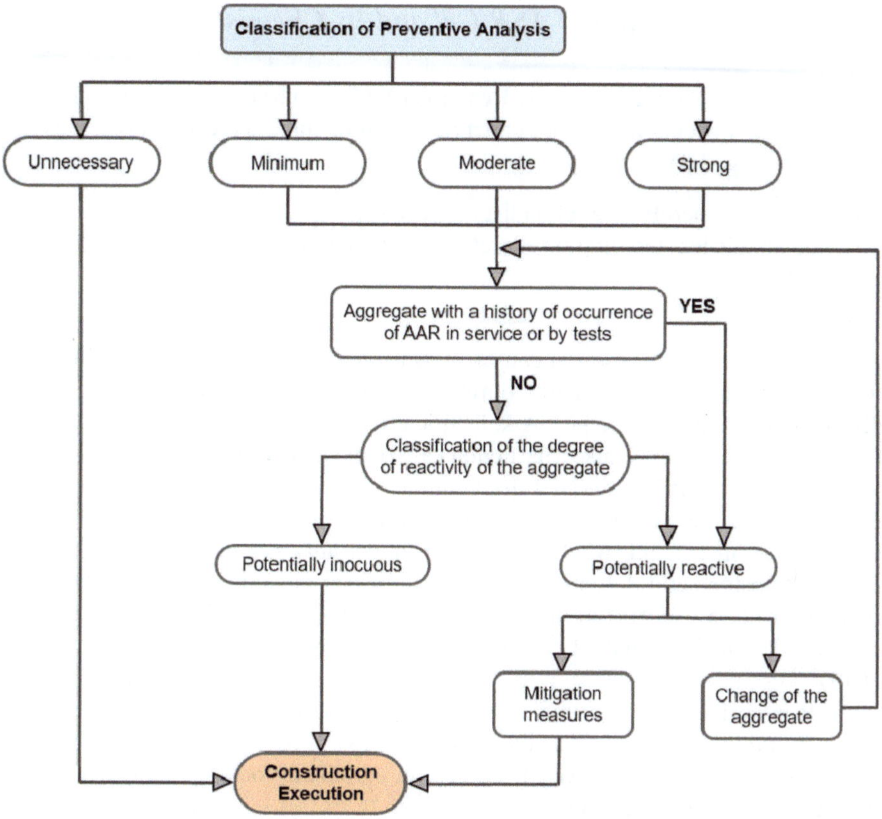

Fig. 2.3 General flowchart for use of aggregate on concrete (adapted from NBR 15577-1:2008)

According to the flowchart, whenever it is necessary to classify the degree of reactivity of an aggregate, petrographic analysis must be carried out, and subsequently, it must be done the reactivity test by MABA—*Método Acelerado em Barras de Argamassa* (Accelerated Mortar Bar Test—NBR 15577-4:2008), which classifies the aggregate as potentially reactive when its expansion limit is equal to or greater than 0.19%, at 30 days of testing. To confirm the results obtained by MABA and/or there is a necessity for more accurate assessment of the reactivity of an aggregate, the use of the MPC—*Método dos Primas de Concreto* (Concrete Prisma Test—NBR 15577-4:2008)—is required. For the aggregate to be considered potentially reactive by the MPC, the expansion limit has to be equal to or greater than 0.04%, at one year of testing. Values below those limits suggest that the aggregate is potentially innocuous. If there is a discrepancy in the results provided by MABA and MPC, it is suggested that only the results on concrete specimens must be used.

2.1.2 *The Formation of ASR Gel and Its Chemistry*

In a general way, Alkali-Aggregate Reaction (AAR) can be considered as a chemical reaction which occurs inside of a concrete structure, and this reaction involves four agents (see Fig. 2.4):

(i) reactive minerals located on the aggregates;
(ii) the alkalis located on the concrete pores moisture;
(iii) water;
(iv) high temperature.

The product of this reaction causes expansion and consequently cracking, drastically interfering on the structure durability (Hasparyk 2011). Commonly, AAR is divided into two different types: Alkali-Silica Reaction (ASR) and Alkali-Carbonate Reaction (ACR). There is also a particular case of the ASR which is the Alkali-Silicate Reaction (ASR). ASR is derived from certain siliceous rocks and minerals, such as opal, chert, microcrystalline quartz, and acidic volcanic glass, present in some aggregates. In the other hand, ACR is derived from certain carbonate rocks, particularly calcitic dolostone and dolomitic limestones, present in some aggregates. However, ASR is far more widespread than ACR, and now is recognized as a major cause of concrete deterioration in the USA and numerous countries worldwide (Thomas et al. 2013).

ASR is derived from certain siliceous rocks and minerals, such as opal, chert, microcrystalline quartz, and acidic volcanic glass present in some aggregates. In the other hand, ACR is derived from certain carbonate rocks, particularly calcitic dolostone and dolomitic limestones.

The ASR process happens in several stages and it's also very complicated to understand and exposed, but the works of Dent Glasser and Kataoka (1981, 1982) proposed that the phenomenon can be summarized in two steps. The first step is

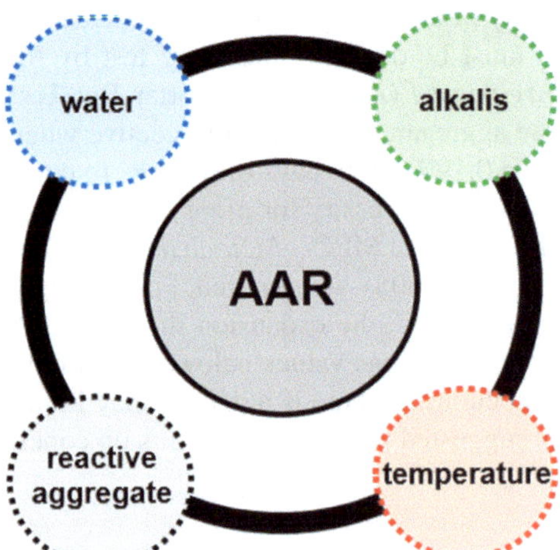

Fig. 2.4 Diagrammatic representation of AAR

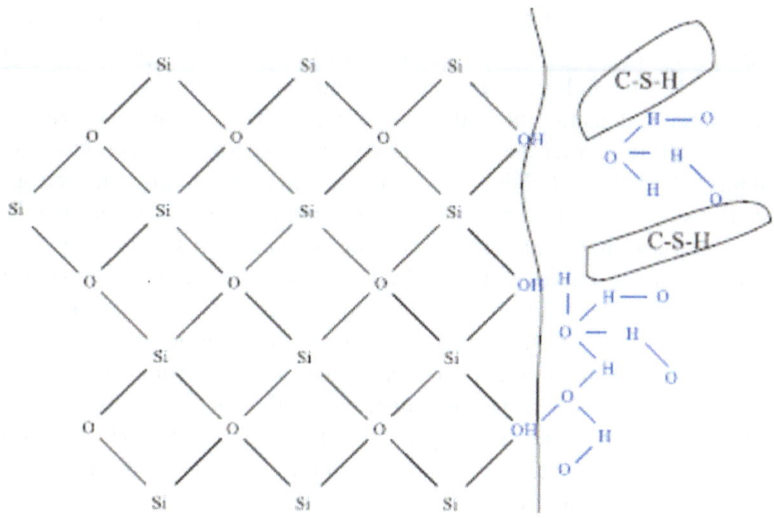

Fig. 2.5 Illustration of aggregate being attacked by alkali hydroxyls ions

the rupture of the aggregate siloxane networks (Si–O–Si) caused by the attack of hydroxyl ions (see Fig. 2.5), which produces alkali silicate and silicic acid, immediately reacting with further hydroxyl ion, and as a result the product of this reaction (the alkali silicate or alkali-silica gel) is amorphous and hygroscopic. The second step is the expansion of the alkali-silica gel by absorption of free water, causing cracking on the cement paste and on the aggregate (depending on the type of the aggregate).

According to Powers and Steinour (1955a), the ASR product is a solid in the absence of water, and this solid can be called basic gel, which has the capacity to imbibe water molecules, generating extensive swelling of the gel and transforming it on a fluid. The authors stated that: "Hydrous silica aggregates, as already mentioned, may of course already contain substantial silanol bonding. These silanol groups react readily with further hydroxyls. As more siloxane bonds are attacked, a gel-like layer forms on the surface of the aggregate. Some silica may even pass into the solution."

The negatively charged species of this gel attract positive charges, which are present in the pore solution in the form of mobile species such as sodium, potassium, and calcium. They diffuse into the gel in sufficient numbers to balance the negatively charged groups.

The presence of these ions determines important properties of the gel. If the gel is formed in an environment rich in Ca_2^+, mainly this species will be taken up by the gel, soon forming particles of C–S–H as a separate constituent. Thus, the gel is transformed into a rigid and unreactive structure. This process may be considered similar to the pozzolanic reaction in concrete. If the pore solution, however, is low in Ca_2^+, the gel takes up mainly Na^+ and K^+. This results in a more viscous consistence of the gel. This gel will also imbibe further water, which will cause its large expansion.

If the gel is in contact with CO_2, which occurs when it is exposed to the atmosphere, it carbonates. This is observed on the surface of cut concrete specimens that have undergone ASR. The ability to imbibe water and ionic species as Na^+ and K^+ cause the water contents, densities, and other physical properties to vary over wide ranges. It is very difficult to predict the species of gel that may form in a certain concrete.

Saouma et al. (2015) detail the ASR in stages, taking as reference the petrographic interpretations on the evolution of the reaction. For the stage I (*Micro-Nucleation*), the reaction starts around the aggregate causing no expansion. On the stage II (*Development*) occurs the formation of the gel in the pores of the reactive aggregates which starts the expansion, and the pressure exerted by this expansion to cracking in the aggregate. On the stage III (*Acceleration*), the process of deterioration becomes more noticeable as the cracking in the reactive aggregate starts to propagate to the cement paste. This cracking grows in size and quantity, and the gel permeates through them, reaching distant voids in the cement paste. For the stage IV (*Deterioration*), severe damage occurs in the structure, for example rupture of the rebar reinforcement and total loss of structural integrity.

2.1.3 Swelling and Cracking Mechanisms Related to ASR

The mechanism of expansion and cracking is still not clearly understood, but is widely accepted that the deleterious process of ASR on concrete structures is due to water absorption by the gel, causing its expansion and increasing internal stress. When the tensile strength of the concrete skeleton is exceeded, the swelling of gel creates and propagates cracking.

The osmotic pressure theory suggests that the cement paste act as a semipermeable membrane when it's surrounding the reactive aggregate, preventing the diffusion of alkali-silicate ions from the reactive sites into the surrounding cement paste, while water is still able to pass through the pore solution. Therefore, an osmotic pressure cell forms and the alkali-silica gel swells with increasing hydrostatic pressure, inducing cracking in the cement paste (Hansen 1944; Dent Glasser 1979; Poole 1992).

The mechanical expansion theory, proposed by Deschenes (2009), suggests that a solid alkali-silicate layer is formed on the reactive aggregate surface. This layer absorbs water from the pore solution and changes from solid to a gel, followed by the absorption of more water by the gel and then it swells, inducing cracking on the cement paste.

Powers and Steinour (1955b) claimed that both of the theories aforementioned—the osmotic pressure theory and mechanical expansion theory—are fundamentally the same. The amount of absorbed water strongly influences on the forms of the alkali-silica gel, and it can be solid or colloidal fluid. Both phenomena, the osmotic pressure cell of fluid gel and solid gel swelling, primarily generates the expansion of concrete. According to Durand et al. (1992), the gel is capable to absorb in volume of water approximately 200–400% of its weight in a saturated atmosphere. In that case, it's possible to calculate the volume and the mass of the gel formed on the surface of the reactive aggregate as the gel can occupy the porous media.

In addition to the theories shown before, the works of Bažant and Steffens (2000) and Dron and Brivot (1993) assumed that the expansion is caused by the swelling pressure accumulated in the Interfacial Transition Zone (ITZ), and the dissolved silica is diffused far away from the reactive aggregates into the connected porous matrix of the cement paste. A pressure in the gel is generated by the imbibition of water, which is initially released by pushing the gel to permeate the micropores in the cement paste located near the surface of the aggregate. The pores around the reactive aggregate and inside the cement paste get fully filled by the gel, and further formation of even more gel induces the increasing internal pressure, leading to cracking of the cement paste.

Although the aforementioned works affirm that the cracking only occurs in the cement paste and not on the aggregate, the researchers Ponce and Batic (2006) found that cracking patterns of ASR-affected concrete depend on the mineralogical nature of aggregate. Those researchers carried out the petrographic examination in aggregates with gel formation and expansion at the aggregate-paste interface, leading to two main conclusions:

- Some aggregates—opal and vitreous volcanic rocks—caused cracking only in the cement paste;
- Mixed mineralogy aggregates—more common in the field—generated cracks in both the cement paste and the aggregates.

On Fig. 2.6 it's shown the two examples listed before. Figure 2.6a shows the cracking on the cement paste within concrete with vitreous volcanic rocks on its mixture. Figure 2.6b shows the cracking in both cement paste and siliceous aggregate.

Some researchers validated the conclusions of Ponce and Batic (2006), and they stated that when hydrate gel starts to be formed inside the reactive aggregate, it develops an expansive pressure causing tensile stress in the aggregate, and such

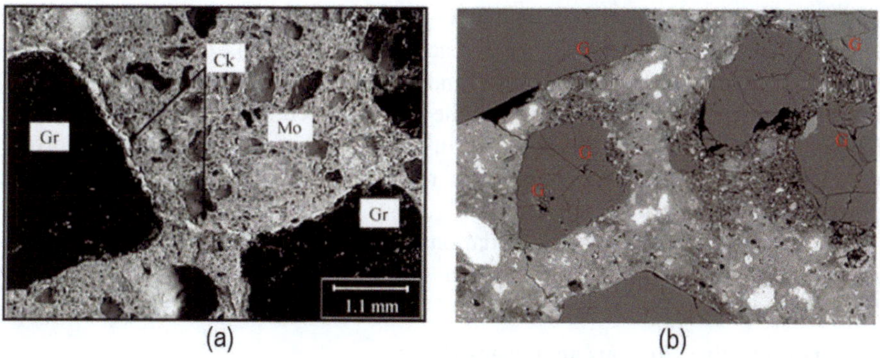

Fig. 2.6 Crack patterns due to different types of aggregates within ASR-affected concrete

pressure is finally released by cracking both the aggregate and the surrounding cement paste (Idorn 2001; Garcia-Diaz et al. 2006).

Other cracking mechanism was proposed by Ichikawa and Miura (2007), suggesting that the diffusion of the viscous alkali-silicate gel doesn't happen because the reactive aggregate is tightly packed with an insoluble and rigid reaction rim, although it allows the alkaline solution to penetrate. After the formation of the rim, the excess of alkali-silica gel generated is stored inside the aggregate, thereby generating expansive pressure and when the strength of material is exceeded, the cracking of the aggregate and the surrounding cement paste occurs.

2.1.4 Factors Influencing on AAR

To better understand the AAR phenomena, it is important to highlight the main factors which have influence on the reaction. As mentioned before, for AAR to occur is needed the presence of reactive aggregates, alkalis on the pore moisture in the cement paste, sufficient humidity and high temperature.

But there are also other factors which play significant role on the expansion due to AAR, such as the porosity of the cement paste and confining stresses on the concrete structure. Therefore, those factors can be divided in three groups:

(i) influence of the concrete composition;
(ii) influence of the environment;
(iii) influence of the confining stresses (Morenon 2017; Balabuch 2018).

2.1.4.1 Influence of the Concrete Composition

Type of Aggregate

The rocks which contain potential reactive silica can react in different velocities, and as a consequence of this variation, the aggregates can be classified by the alkaline reactivity as "fast reactive" to "normal reactive" (5–20 years), "slow reactive" (more than 15–20 years) and "non-reactive" (Lindgård et al. 2010). The aggregate reactivity is directly linked to the velocity which the reaction occurs as well as the type of the reaction. The increase on the permeability of the aggregates with higher porosity may lead to an increase in the aggregate reactivity (Zhang et al. 1999; Broekmans 2002).

According to LCPC recommendations (LCPC 1994; FD P18-464—*Béton—Dispositions pous prévenir les phénomènes d'alcali-réaction*, 2014), aggregates can be classified into three groups:

(i) non-reactive siliceous aggregates (NR);
(ii) potentially reactive aggregates (PR);
(iii) potentially reactive aggregates but with a pessimum effect (PRP).

Indeed, for some aggregates a maximum swelling can be observed depending on the content of the reactive aggregate (Hawlett 2003). The peak of the pessimum expansion varies according to the aggregates used. The kinetics of the reactions also strongly depend on the nature of the aggregates.

Tiecher (2006) analyzed 40 samples of Brazilian aggregates, where 30 of them were rocks (granites, rhyolites, basalts, limestone and gneiss) and the other 10 were sand. Using the petrographic analysis, the author verified that all the analyzed aggregates (without exceptions) showed potential reactive mineralogical constitutions. The analyzed aggregates in general had deformed silicate minerals in its matrix, which may lead to the occurrence of AAR.

Granulometry

The specific surface of the aggregates has an influence on the swelling, more gel can migrate into the connected porosity for a small size aggregate than for a larger one (Gao et al. 2013). In addition, a pessimum effect is observed depending on the size of the aggregates (Multon et al. 2010; Gao et al. 2013). Multon et al. (2010) stated that the size of the aggregate which causes the highest AAR expansion depends on the nature and composition of the aggregate. For fast reactive aggregate it was verified that the quantity of soluble silica was similar for the different particle sizes of a certain aggregate, and the expansion varied to different sizes as the higher expansions was verified to coarse particles (Multon et al. 2008).

This phenomenon could be explained by the quantity of silica consumed during the cracking of the aggregate. If the aggregate is small, cracking will take place for an aggregate which silica will be strongly attacked. For a larger aggregate, its core won't have reacted yet at the time of cracking. The produced ASR gel will only cause a small increase in pressure due to the space created by the cracking (Gao et al. 2013).

A scale effect is also observed: the larger the size of the aggregates in relation to the size of the test pieces, lower will be the final swelling. Gao et al. (2013) explains this phenomenon by the production of the gel in the created cracking, as the crack depends on the size ration of the swelling inclusions and the test specimen.

The difference on the kinetics of the reaction between the gravel and the sand is used to determine the advancement of the reaction on testimonial blocks extracted from concrete structures (Grimal 2007).

Alkalis Content

The alkali content provides an increase in the concentration of OH^- ions in the concrete pore moisture, which are responsible for the attack to the silica in the reactive aggregates (Hasparyk 2011). This content must be sufficient to allow the reaction (Poole 1992; Berra et al. 2005).

The Portland cement is the main source of alkalis in the concrete, and higher is the alkalis content in the cement or higher the cement consumption, higher will be the expansions. But the Portland cement is not the only source of alkalis in the concrete, as some minerals in the aggregates (feldspar or mica) can also provide the alkalis (Diamond 1989; Bérubé et al. 2002).

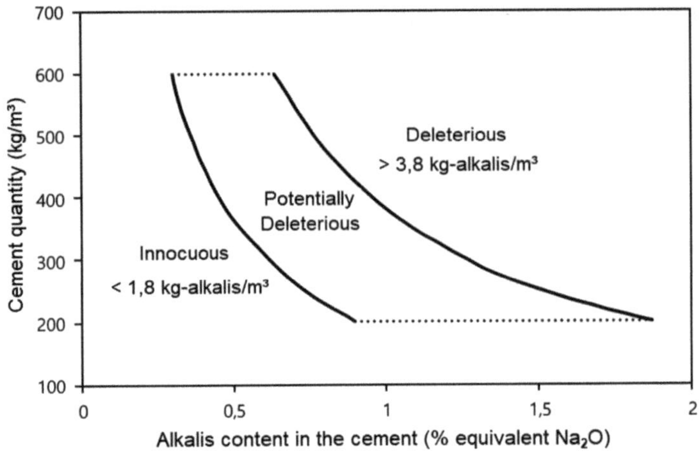

Fig. 2.7 Final expansions of ASR-affected concrete specimens depending on their size and the size of the aggregates (adapted from Oberholster 1983)

Oberholster (1983) proposed some guidelines in terms of cement content in concrete and active alkalis content of cement to evaluate in which conditions a concrete specimen would suffer from ASR-expansion (Fig. 2.7). Posteriorly, in order to avoid the occurrence of AAR, it was stablished a limit in the alkali content of 0.6% of equivalent Na_2O ($Na_2O + 0.658K_2O$) (NBR 15577-1:2008; Mehta and Monteiro 2014).

The leaching of alkalis is an important phenomenon to take into account in testes with small concrete specimens (Lindgård et al. 2013). In fact, Multon and Sellier (2016) specified that a competition between alkali leaching and the diffusion of alkalis in the aggregates can participate in the scale effect observed in the case of test specimens stored in high humidity conditions.

Porosity of the Cement Paste

A high value of porosity of the paste plays two confronting roles when related to AAR. Greater is the porosity, greater is the volume of available moisture on the pores and greater is the space which the gel has to permeate without causing damage to the concrete, because the available voids would be larger, although the diffusion would be accelerated.

In the other hand, lower is the porosity, lower will be the velocity of the AAR, delaying the advancement of the reaction, because the gel and the water would be restricted in their mobility, causing a reduction on the expansion of the gel and even preventing the formation of the gel. Jensen (2004) show a technique to reduce the pessimum expansion using air-entraining in the concrete mixture, and for values of 4% of air insertion there is a reduction about 40% of the pessimum expansion, although the technique of air-entraining can drastically reduce the structural properties (e.g., compression strength) of concrete (Silva et al. 2009; Wawrzeczyk and Molendowska 2017).

The impact of the permeability of the gel (in the cement paste) on the expansions before cracking is limited to few cases of very reactive aggregates. In most cases, significant expansions are generated before the detection of the gel in the cement paste (Sanchez et al. 2015). This phenomenon occurs because once AAR has begun, it rapidly generates an intra-porous stress (Sellier et al. 1995; Capra 1997).

2.1.4.2 Influence of the Environment

Temperature

The temperature is directly linked to the advancement of the reaction: higher is the temperature, faster is the reaction. This effect is simulated on the laboratory tests to accelerate the AAR which would take years in the room temperature (Diamond et al. 1981; Larive 1997).

As the temperature is increased, the solubility of the silica is also increased and also is perceived a decrease on the solubility of the calcium hydroxide. Making the silica more suitable for reaction and decreasing the capacity of the calcium hydroxide to combine with the gel, is obtained an increase in the velocity of the expansion by AAR (Collins and Bareham 1987).

The effect of the temperature on the kinetics of the reaction can be modeled with an Arrhenius law (Eq. 2.1), where the term "Ea" is the activation energy, which varies from 40 and 53 kJ/mol, and the term "R" is the perfect gas constant, which the value is equal to 8.1314 J/(mol K) (Larive 1997; Ulm et al. 2000; Poyet 2003; Grimal 2007).

$$\varepsilon_{AAR}^{calcul} = \exp\exp\left[-\frac{Ea}{R}\cdot\left(\frac{1}{T_{ref}}-\frac{1}{T_{calcul}}\right)\right]\cdot\varepsilon_{AAR}^{ref} \tag{2.1}$$

Lombardi et al. (1997) stated that the increase in the temperature promotes an increase in the number of moles of gel created, and Furusawa et al. (1994) also affirms that the increase in temperature causes a change in the coefficient of apparent diffusion in the aggregates.

Larive (1997) stated that apparently the swelling asymptote is not modified by the temperature, although other researchers (Diamond et al. 1981; Wood et al. 1987) showed results that confront this point. Wood et al. (1987) even found swellings 4 times greater at low temperatures (around 13 °C) than at high temperatures (around 38 °C). Disturbing effects may had occurred on these experimental tests, such as alkalis leaching, gel permeation in cracks, the chemistry of the gel and even thermal damage.

The concrete expansion caused by ASR shows picks at temperatures close to 40 °C, and this phenomenon might be related to the lower viscosity of the gel for temperatures higher than 40 °C, which would result in the decreasing of the expansion as the gel can permeate more easily through the pores of the cement matrix (Dyer 2015).

Relative Humidity

The water inside the concrete pores is also considered as one of the main factors to the developing of AAR, because the expansion induced by ASR is a consequence of the water imbibition by the gel and therefore the extension of the expansion depends on the amount of available water in the pores. Basically, the water acts in three different ways:

(i) on the transportation of alkali ions in the interstitial solution to the reaction sites by diffusion;
(ii) on the silica dissolution;
(iii) on the expansive mechanism of the gel (Mehta and Monteiro 2014).

The ASR gel is hydrophilic which absorbs a lot of water molecules. Some experimental data showed that the expansion was directly linked to the relative humidity (RH) of the environment (Larive 1997; Poyet 2003; Multon 2003). Although the modeling of this effect became more precise when the RH was replaced by the saturation degree of the concrete.

Poyet (2003) shows that is a threshold at approximately 70% of relative humidity which above this value the swelling is no longer negligible. The final swelling seems less important for a RH of 100% than for the value of 96%, and this phenomenon can be due to the alkalis leaching. Larive (1997) states that a late addition of water can accelerate the kinetics of the reaction. According to Bazănt and Steffens (2000), for ASR takes place in the concrete, is necessary a RH of 85%. To Léger et al. (1996) this value is around 75%.

The water is an important factor on the process of hardening and curing of concrete, which are used water/cement ratios varying from 0.35 to 0.60 to conventional concrete. Therefore, the humidity resulting from the water used on the concrete production (found on the pores of hardened concrete) is enough to react, contributing to the development of AAR (Mehta and Monteiro 2014).

2.1.4.3 Influence of the Confining Stresses

The confining stresses are considered as one of the most important factors which influences on the expansion induced by AAR, because they can limit the free expansion of the concrete structure or even prevent it. A compression stress causes a reduction in the value of the expansion on the direction which the stress is applied, and in the same way, a tensile stress can increase the expansion on the direction which the stress is applied (Pourbehi et al. 2018).

Capra and Bournazel (1998) showed that when a compression stress is applied in one of the main directions, the expansion in the orthogonal direction of the applied stress is a sum of the expansion that would occur normally in this direction and the expansion related to the restricted direction, where the direction which the stress is applied is the direction 1. The curve of free expansion was calculated as a sum of the expansions on the directions 1 and 2 in the free stress condition.

2.1.5 Numerical Modelling of ASR

Numerical models able to describe the AAR have been developed since the 1980s (Carasek et al. 2016), but the scientific community hasn't yet reached a consensus on a model able to fully describe and reproduce the effect of the evolution of AAR-affected concrete structures, nor for which parameters should be considered in the model (Gameleira et al. 2016).

In literature, there is a several number of mathematical models to represent the expansion of ASR-affected concrete structures, and it's because of the complexity of this phenomenon: the heterogeneity of concrete, random distribution of reactive sites, dependence on the environmental conditions and a large number of theories to describe the mechanisms associated to the reaction (Capra and Sellier 2003; Pignatelli 2012). In addition, the complexity of simulating AAR using numerical methods gets even higher because of the large amount of physical phenomena to consider, such as concrete reactivity, thermal activation, moisture (depends on the rheology of concrete) and the damage interaction (Grimal et al. 2007).

According to Bazănt et al. (2000), to perform an ASR-affected concrete structure simulation, some basics aspects must be modelled first and then combined:

- Chemical reaction kinetics and diffusion process: determines the degree of ASR and potential expansion;
- Mechanical fracture, expansion and deterioration of concrete structures: describes the material and structural degradation.

One of the difficulties to simulate AAR-affected concrete relies on the fact that the concrete is a very complex heterogeneous material which has random microscopic structure, with different scales of length. In the macroscopic scale, the concrete is considered as a homogeneous material. In the mesoscopic scale, there is the inclusion of the cement matrix, aggregates, pores (voids) and the ITZ, with sufficient length distribution. The microscopic scale is the finest structural scale, represented by the microstructure of the hardened cement paste, composed by the hydration products, micropores and non-hydrated residual clinker (Dunant and Scrivener 2010; Nguyen et al. 2012).

According to Pan et al. (2012), the existing mathematical models can be classified based in its scale:

- Microscopic Models: Those models deal with the chemical reactions only and the theoretical approach generally examines a Representative Volume Element (RVE), which consists of a cement-aggregate system. They were widely developed to predict the pessimum size of aggregate, and such size is responsible for generating the maximum expansion of ASR gel in affected concrete structures.

- Mesoscopic Models: Those models represent the mechanism of mechanical deterioration of ASR in a material scale, proposing to establish an interface with large structures. Anisotropy in concrete is expressed through the coupling of the multiphase of concrete (aggregate, cement paste, pores and ASR gel). ASR swelling is represented as a dilatation of gels randomly distributed or an equivalent expansion of aggregate particles.
- Macroscopic Models: Those models deal with the coupled chemo-mechanical behavior of ASR-induced expansion of the affected concrete structures, combining the chemical reaction kinetics with linear or non-linear mechanical constitutive, formulated in the framework of Finite Element Method (FEM). This type of model aims to reproduce and predict the long-term behavior observed in real AAR-affected concrete structures.

Later, the review work of Esposito and Hendriks (2019) brings a new classification for the available AAR-models, categorizing them based on their starting scale, which is defined as the level at which the input parameters are defined. The authors summarized as follows:

> The chemical process involves silica ions, available in the aggregates, and alkali ions, mainly present in pore solution together with water (reaction products level). The formed alkali-silica gel, when exposed to moisture, tends to swell. Its expansion, while confined in the pore structure of concrete, builds up an internal pressure with the consequent formation of cracks in the aggregates and in the cement paste (aggregate level). As a consequence, the concrete is expanding and its mechanical properties are degrading (concrete level). The reduction in material resistance compromises the performance of the structure in terms of both capacity and durability (structural level).

Several modelling approaches were developed with different aims as the AAR is considered a multiscale phenomenon, which takes place at the reaction products level and may implies consequences up to the structural level, this reaction had been studied by experts from different areas (e.g., structural engineers, material engineers, geologists) in the main fields of material sciences and structural mechanics (Esposito and Hendriks 2019).

- Models based on concrete expansion (concrete level): Those models focus on the analysis of massive structures (e.g., concrete arch-dams). An imposed expansion strain in concrete is used to simulate the effects of ASR, and the evolution of this strain over time was derived from reaction kinetics laws (based on thermodynamic principles) and phenomenological laws linked to the environmental conditions. Examples of this type of modelling can be found in the works of Larive (1997), Léger et al. (1996), Capra and Bournazel (1998), Ulm et al. (2000) and Comi et al. (2009).
- Models based on internal pressure (aggregate level): Those models were developed thanks to the technological advancement in microscopic investigation on the ASR mechanisms at aggregate level, performed at the beginning of this century. The concrete was represented as a heterogeneous material composed by coarse and fine aggregates, cement paste and the ITZ, defining the microstructure of concrete and the mechanical properties of each phase. The gel swelling was direct related

to the expansion of concrete. Example of this type of modelling can be found on the works of Bazănt et al. (2000), Dormieux et al. (2004), Comby-Peyrot et al. (2009), Wu et al. (2014) and Esposito and Hendriks (2016).

- Models based on gel production (reaction products level): Those models were developed at reaction products level with the aim to couple the physical chemistry of ASR with the mechanical behavior of concrete. The kinetics of the reaction were expressed as a function of the change in volume or mass of the ASR gel, and this change in volume/mass was translated into an imposed pressure or strain at aggregate level. Some models were developed in the framework of micro-poro-mechanical theory to investigate the connection between the swelling products at aggregate level and the consequent concrete expansion. Examples of this type of modelling can be found on the works of Ulm et al. (2000), Grimal et al. (2007), Dunant and Scrivener (2010), Pignatelli et al. (2013) and Charpin and Ehrlacher (2014).
- Models based on ions diffusion–reaction (reaction products level): Those models were developed based on the chemical process translated as ions diffusion–reaction mechanisms, at reaction products level, and they aim to explain why and how the chemical reaction leads to the formation of ASR gel. The phenomena of pessimum size effect and the alkali leaching could be explained by postulating the diffusion of alkali ions via the pore solution and the following chemical reaction. Examples of this type of modelling can be found on the works of Bazănt and Steffens (2000), Suwito et al. (2002), Poyet et al. (2006), Multon et al. (2009), Sanchez et al. (2014), Nguyen et al. (2014), Multon and Sellier (2016) and Morenon (2017).

As mentioned before, this work will implement the model presented in the last technical report provided by the LMDC (Sellier 2018), which was firstly developed by Morenon (2017). The chemical model is based on the pressure exerted by the generated volume of ASR gel based in the poro-mechanical framework. Such model is described with more details on the next chapter (Chap. 3), but to understand its main assumptions and hypothesis, is necessary to give to the reader a brief review on the models based on the generated volume of gel.

The models responsible for describing the mechanical behavior of the swelling in concrete structures are discussed in this chapter, in Sect. 2.2.4, after a brief review of the main mechanical behaviors presented on concrete structures.

2.1.5.1 Modelling of the Generated Volume of ASR Gel

Some models were developed based on the diffusion of ions in a Representative Volume Element (REV) which incorporates the aggregate (Furusawa et al. 1994; Bazănt and Steffens 2000; Poyet 2003).

As one of the pioneer's works which applied the concept of REV in ASR modelling, Furusawa et al. (1994) developed a model based on the diffusion of alkali ions and hydroxides in the aggregate, and such model was able to describe a porous area around the aggregate which becomes filled with the generated volume of

ASR gel. The expansion takes place after the vessel gets fulfilled by the gel, and the volume of this vessel varies according to how much of reactive silica was consumed.

The mathematical model for the kinetics of ASR, presented by Bazănt and Steffens (2000), was developed aiming to predict the pessimum size of ground waste glass aggregate. It was a simplified approach on the complex chemical reactions and micro-diffusion processes with a good agreement with the experimental tests results, and the ingress of water into the concrete is modelled by a classical Fick law. Although the model was not available for regular concrete (e.g. crushed stones as aggregates) and the fracture mechanics were not taken into account (this model was developed separately in Bazănt et al. (2000)).

Nonetheless, the assumption is that the species migrate towards the aggregate, therefore the silica is dissolved and reacts. A gel is formed around the aggregate, subsequently it's filled with water and the expansion is generated. Water is imbibed by the gel through diffusion. The water content in the gel (ξ_w) is then calculated as (Eq. 2.2):

$$\xi_w(x,t) = w_s \cdot \left(\frac{1 - \frac{z(t)}{x}}{1 - \frac{2z(t)}{D}} \right) \tag{2.2}$$

where w_s is the water content around the aggregate; $z(t)$ is the radius of the aggregate that remained unreacted; D is the initial diameter of the aggregate.

It was assumed that two molecules of water are needed to dissolve a reactive silica molecule. Therefore, the equation to calculate the mass concentration of the gel (ξ_g) is then expressed as (Eq. 2.3):

$$\xi_g = \frac{4\pi}{3} | \left(\frac{D^3}{8} - z^3 \right) \cdot \rho_s \cdot \frac{m_g}{m_s} \cdot \frac{1}{s^3} \tag{2.3}$$

where ρ_s is the mass density of the reactive silica; m_g is the molar mass of the ASR gel; m_s is the molar mass of the reactive silica; s is the number of moles of the reactive silica.

The generated gel will then combine with water, when the RH is equal or higher than 85%, resulting in a swelling product. The water imbibition by the gel is expressed as an Ordinary Differential Equation (ODE) and can be written as (Eq. 2.4):

$$\frac{\partial w_i}{\partial t} = \dot{w}_i = \frac{n(P) \cdot m(RH) \cdot \xi_g - w_i}{\left(\frac{\delta_c^2}{C_{w_i}} \right)} \tag{2.4}$$

where w_i is the mass of water imbibed by the gel; δ_c is the thickness of the area where the diffusion takes place; C_{w_i} represents the diffusivity of water into the gel; $n(P)$ is the empirical function to represent the decrease of "$\dot{w}_i$" with the increase of pressure; $m(RH)$ is empirical function to represent the decrease of water imbibition by the gel with the increase of RH.

Poyet (2003) assumed that the aggregates can be represented by spheres (REV) and it is attacked by different chemical species (Na^{+}, K^{+}, Ca^{2+}), also the granular distribution directly impacts the kinetics of the reaction and the amplitude of the swelling. The mass conservation equation (Eq. 2.5) for any of the aforementioned species (specie X) makes it possible to manage the degradation of the aggregate by a diffusion mechanism. From this equation, written for any chemical compound X, the volume of gel is calculated and the subsequent swelling is evaluated by an energetic method which takes into account the volume of the connected porosity.

$$\frac{\partial}{\partial t}\left(\underline{p_g}\cdot\underline{Sr_g}\cdot\underline{X_g}\right) = -div\left[\underline{D_g^x}\cdot\overrightarrow{grad}\left(\underline{X_g}\right)\right] + \underline{S_g}(X) \tag{2.5}$$

where $\underline{p_g}$ is the porosity of the aggregate; $\underline{Sr_g}$ is the degree of saturation of the aggregate; X is the concentration of the chemical specie X; $\underline{D_g^x}$ is the coefficient of diffusion; $\underline{S_g}$ is the coefficient to permit the consumption of the chemical species by the reaction (at the exterior of the aggregate, this term corresponds to the consumption of the specie X into other aggregates).

With the aim to enhance the model developed by Poyet (2003), Multon et al. (2009) added a kinetic law for the alkalis fixation. Therefore, the swelling may be influenced by the diffusion or alkalis fixation. This model integrated to a finite element code (FEC) was able to reproduce the experimental data obtained in concrete specimens subjected to alkali gradient by external leaching.

The most used model to reproduce the kinetics of the ASR gel swelling is the empirical model proposed by Larive (1998), which the advancement of the reaction (ξ) is used to assess the volume of gel created over time (Eq. 2.6). The main assumptions are that the ASR gel is considered as an incompressible fluid and the expansion is proportional to the advancement of the reaction, the term ξ is then multiplied by the final deformation to calculate the deformation for any given time (t). The term τl is the latency time which is responsible for managing the start of the swelling. The term τc is the characteristic time which is responsible for defining the kinetics of the reaction between the beginning and the end of the swelling, establishing the asymptote.

$$\xi(t) = \frac{1 + e^{-\frac{t}{\tau c}}}{1 + e^{-\frac{t-\tau l}{\tau c}}} \tag{2.6}$$

This type of curve "S" is typical of the internal swelling reactions. However, the parameters cannot be linked to measurable physical properties.

A differential law for the advancement of the reaction was proposed by Grimal et al. (2018). The advancement was calculated (Eq. 2.7) based on the temperature T (Arrhenius law) and the saturation degree Sr (Poyet et al. 2006). The advancement A goes from 0 (reaction not started) to 1 (reaction totally finished). Then, the advancement of the reaction is multiplied by a maximum volume of ASR gel that can be created (calibrated by the user regarding the expansion curves), allowing to obtain the volume of ASR gel for any given time.

$$\frac{\delta A}{\delta t} = \alpha_0 \cdot e^{\frac{Ea}{R}\left(\frac{1}{T_{ref}} - \frac{1}{T}\right)} \cdot \left(\frac{\langle Sr - Sr^0 \rangle^+}{1 - Sr^0}\right) \cdot \langle Sr - A \rangle^+ \tag{2.7}$$

where α_0 is a parameter to calibrate the kinetics; Ea is the ASR activation energy (47,000 J/mol/K); R is the perfect gas constant (8314 J/mol); T_{ref} is the reference temperature (K) of the experimental test; $\langle \rangle^+$ represents the Macaulay brackets, which are the positive part of the terms inside.

The majority of those presented models needs the previous knowledge on the free expansion of a concrete specimen in order to calibrate the parameters which cannot be directly measured. In general, three parameters are necessary to reproduce the curves admitted as representative of ASR-expansion.

In Chap. 3, it's presented the chemical model proposed by Morenon (2017) which follows the work of Grimal et al. (2018), using the advancement of the reaction (A) to generate the volume of gel. This volume exerts a pressure that is computed in the poro-mechanical framework.

2.2 Mechanical Behavior of Concrete

In this topic is presented a brief theoretical review about the concepts linked to creep, shrinkage, poro-mechanical theory and the main mechanical models for AAR-affected concrete structures. The factors which influence the occurrence of those phenomena are commented as well. This explanation is necessary in order to better understand the numerical models and its parameters presented on Chap. 3.

2.2.1 *Creep and Shrinkage*

Time-dependent strains of concrete—at constant temperature—may be subdivided into a load-independent part, called shrinkage, and a load-dependent part, called creep (Bazănt and Panula 1978). The total deformation related to these phenomena is the sum of the strain related to shrinkage and the strains related to creep (Fig. 2.8).

A concrete creep model to be relevant in structural applications must consider several important phenomena, such as non-linearity, multi-axiality, hydration, and thermal and drying effects (Sellier et al. 2016). In order to consider the long-term mechanical behavior of concrete presenting delayed strains is necessary to split the total delayed strain into five different strains: basic creep, drying creep, plastic shrinkage, autogenous shrinkage and drying shrinkage. A poro-mechanical framework used on FEA softwares may be very useful to develop these type of numerical modelling, because it can distinguish the role of water and the constitutive behavior of solid phases.

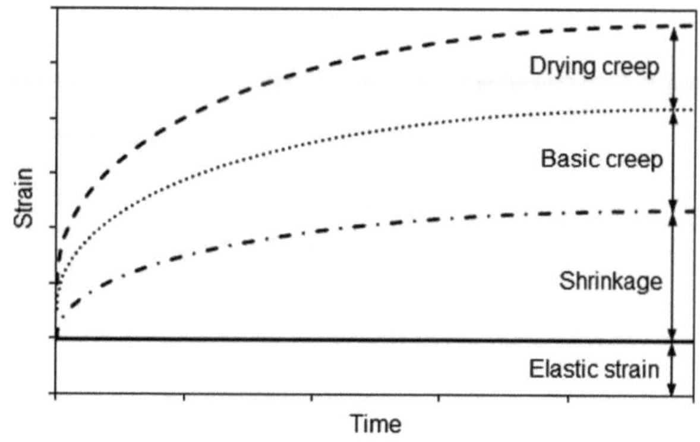

Fig. 2.8 Total deformation of concrete subjected to constant loading (adapted from Balabuch 2018)

2.2.1.1 Creep

A perfect elastic material is the one which the deformation appears and disappears immediately after the application or removal of loading. It's important to note that an elastic material does not imply in a linear relation on the stress x strain graphic. The elastic characteristics of a certain material are a measure of its stiffness.

Rüsch et al. (1983) stated that the velocity at which the load is applied also influences the instantaneous strain recorded on the experimental test. The strain which occurs immediately or concomitantly to the load application is considered elastic, and its later increase when the load application is constant is treated as a slow deformation, also known as creep. There are two types of creep: basic and drying creep. The total creep is calculated as the sum of the basic and drying creep.

When the drying shrinkage is neglected, the basic creep can be taken as the increase in the strain over time, with an applied constant load, under conditions of 100% of RH and no water flux between the concrete and the environment. The drying creep is an additional type of creep which occurs when the concrete is under loading and it's simultaneously exposed to an environment with low RH.

Sellier et al. (2016) stated that at a microscopic scale, the fundamental cause of creep is the constitutive properties of hydration products, generating a viscous behavior due to the instability of the bond between C–S–H sheets. At a macroscopic scale, the cause of creep can be related to other non-linear phenomena such as water content variation and micro cracking of anhydrous phases.

To highlight the importance of such deformation, Kuperman (2007) showed the results of an experimental test where concrete specimens with cement consumption of 350 kg/cm^3 were subject to a constant loading of 10 MPa for 7 days, under RH of 50%. In 4 days, the concrete specimens showed total strains almost 3 times higher than the elastic strain, and at the end of 3 years, the total strain was more than 6 times higher.

In concrete structures, creep is capable of reducing the internal stresses due to non-uniform or restrained shrinkage. On the other hand, considering a huge volume

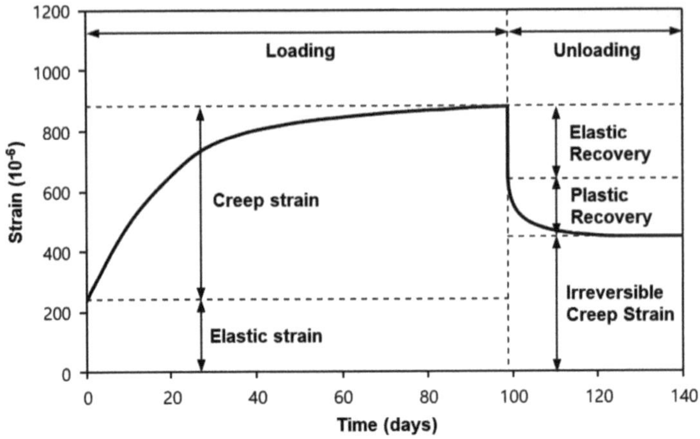

Fig. 2.9 Creep reversibility (adapted from Mehta and Monteiro 2014)

of concrete, the creep alone may cause cracking when the restrained concrete suffers a temperature changing cycle caused by the hydration heat and the subsequent cooling.

Mehta and Monteiro (2014) illustrated the phenomenon of reversible creep of concrete when the loading is ceased. In Fig. 2.9 is observed a concrete specimen subjected to constant loading for 100 days, followed by the removal of the load which lasts for 40 days. After the instantaneous elastic strain, the deformation keeps rising over time as the loading is maintained. When the loading is removed, the instantaneous elastic recovery has approximately the same order of magnitude as the instantaneous elastic strain (for the first application of the loading). As time goes by, there is a gradual reduction in the strain, known as creep recovery or plastic recovery. This plastic recovery occurs faster than the creep, but it eventually stops, leaving a permanent deformation (irreversible creep).

2.2.1.2 Shrinkage

Sellier et al. (2016) stated that the water has complex effects in porous media because it can be adsorbed on porous walls, react with the solid skeleton, and exert capillary and disjoining forces on the solid skeleton. In concrete, the water molecules are found in nanoscopic inter-sheets, causing repulsive or attractive forces depending on the environmental conditions (temperature, RH and mechanical loading), but such effects have several consequences in macroscopic scale.

In a resumed way, shrinkage is caused by the water capillary pressure which exerts a tension perpendicular to the pore walls, and it is balanced by orthogonal compressions forces which are parallel to the pore walls. This water capillary pressure may occur only in a high RH environment, but other forces, like disjoining force variations (occurring in a low RH environment), can have similar consequences on drying cement pastes (Sellier et al. 2016).

Shrinkage can be divided in three: plastic shrinkage, autogenous shrinkage and drying shrinkage. The first occurs when the concrete is on the curing process, and the second one occurs when the concrete is already hardened (Balabuch 2018).

Plastic Shrinkage

According to Dyer (2015), during the process of curing of a concrete structure, it can be observed two simultaneous processes happening on the surface of the concrete (Fig. 2.10a). Firstly, the solid constituents have the tendency to settle which leads to the formation of a water layer on the surface of the concrete (Fig. 2.10b). Secondly, the free water on the surface gets evaporated, leading to the plastic shrinkage when the evaporation rate overcomes the transpiration rate. This happens because the cement particles on the surface of the concrete are no more covered by water, inducing tensile stresses on the superficial layer which is restrained by internal layers of the concrete, leading to the formation of water meniscus between the cement particles (Fig. 2.10c). With the progressive water evaporation, the radius of the water meniscus gets shorter, and for a certain instant of time, the radius are going to be so small that they won't fulfill the space between the cement particles in such a way that the water surface will need to rearrange in a point below the initial surface, leading to air penetration (Fig. 2.10d). Higher is the evaporation velocity, higher will be the plastic shrinkage.

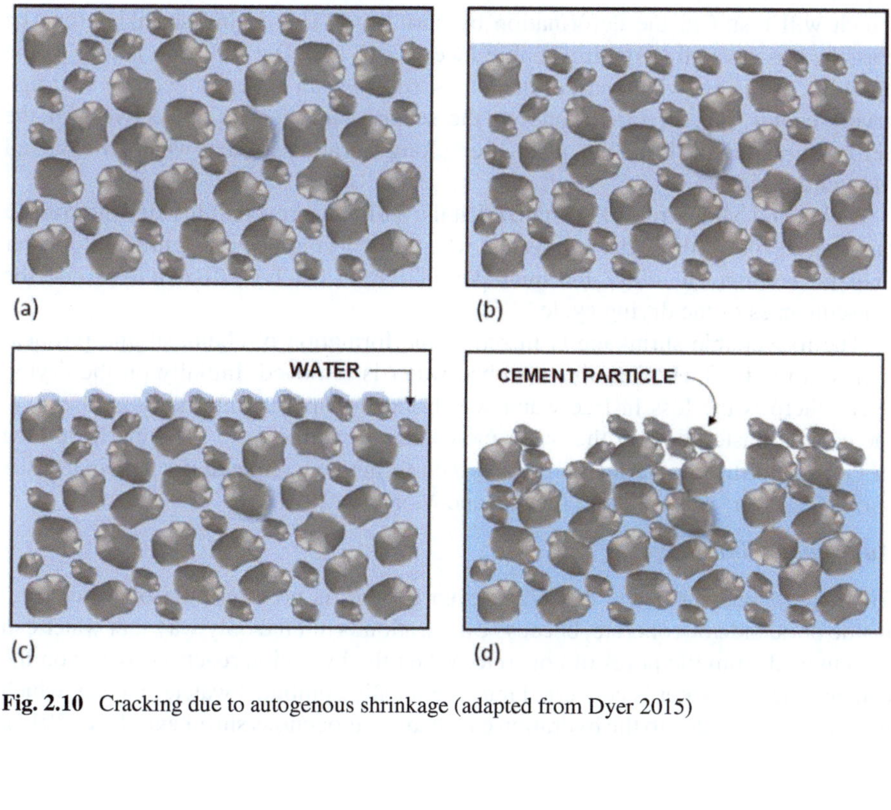

Fig. 2.10 Cracking due to autogenous shrinkage (adapted from Dyer 2015)

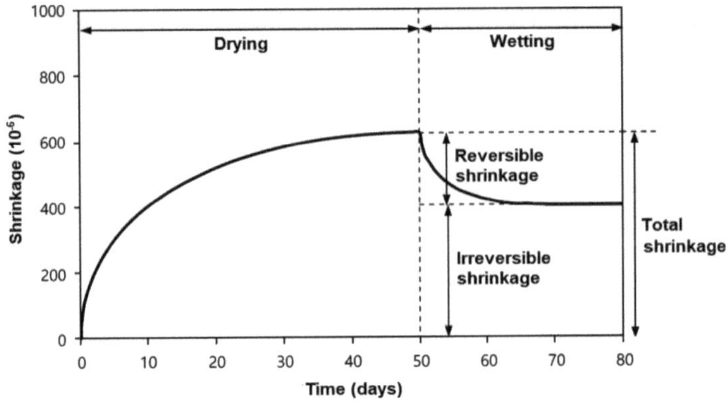

Fig. 2.11 Drying shrinkage reversibility (adapted from Mehta and Monteiro 2014)

Drying Shrinkage

The drying shrinkage occurs in the hardened concrete due to the loss of the physically adsorbed water from the C–S–H, which represents 50–60% of the total volume of solids in an entirely hydrated cement paste. When the saturated cement paste is exposed to lower environmental humidity, it will lead to a dimensional instability which will result in the deformation by shrinkage. The drying shrinkage can be classified as reversible or irreversible (see Fig. 2.11). The first one is due to the humidity diffusion, which part of the total shrinkage can be reproduced by wetting–drying cycles. On the other hand, at the second one there is no recovery of the deformation due to shrinkage, even after a long storage period in water (Mehta and Monteiro 2014).

Mehta and Monteiro (2014) stated that the shrinkage reversibility depends on the duration of the wetting–drying cycles, where the drying occurs significantly faster than the wetting. Therefore, the small period of rains can be sufficient to reverse the consequences of the drying cycle.

The irreversible shrinkage is linked to the formation of chemical and physical bonds in the C–S–H when the adsorbed water is removed. Initially on the drying cycle, there is the loss in free water which induces internal humidity gradients in the cement paste, moving the water molecules over time from the superficial area of C–S–H to the capillaries pores, and then to the concrete exterior leading to the shrinkage of the cement paste (Mehta and Monteiro 2014).

Autogenous Shrinkage

The autogenous shrinkage can occur even if there is no migration of water from inside or outside of concrete, because evaporation is not the only way that water can be removed from the pores of concrete. When the hydration reactions occur on the cement, the free water is converted into chemically combined water, in a way which this loss of water due to the hydration can lead to autogenous shrinkage (Dyer 2015).

This type of shrinkage consists of the macroscopic reduction of the volume of the cementitious materials which occurs when the cement is hydrated after the initial setting time. The autogenous shrinkage occurs in many different types of concrete and is considered as almost negligible for concrete with conventional strength, but it's very important for high-strength concretes and concrete mass. This phenomenon is not related to volume variation due to loss or gain of substances, variation on temperature, application of mechanical loading or restrictions to the structure.

2.2.1.3 Factors Influencing Shrinkage and Creep

In the literature (Troxell et al. 1958; Brooks 1989; Zhang and Malhotra 1995; Mazloom et al. 2004; Ayub et al. 2014; Wendner et al. 2015; Joshaghani et al. 2018), the main factors influencing shrinkage and creep can be divided in 4 main groups:

- Concrete composition: water/cement ratio, type of cement, type of aggregate, admixtures and additives;
- Environmental conditions: relative humidity and temperature;
- Concrete characteristics: element geometry and age of the concrete;
- Mechanical loading: intensity and age of the loading.

Water/Cement Ratio

Wendner et al. (2015) affirms that the w/c ratio plays a major role on the porosity of the cement paste and subsequently on the resistance properties of the concrete. Lower is the w/c ratio, higher will be the concrete resistance. Considering a constant cement paste content, an increase in the w/c ratio will result in an increase of the creep strain.

According to Neville and Brooks (2013), the w/c ratio also influences the quality of the cement paste, which may impact on the shrinkage intensity. Higher is the w/c ratio, higher will be the shrinkage. Considering a certain aggregate content, the shrinkage will be a function of the w/c ratio.

Type of Cement

The hydration of the matrix of the cement paste is influenced by the type of cement used in the mixture, also the creep is influenced by the type of cement as it impacts on the strength of a concrete structure under loading. When the concrete which uses conventional Portland cement is subjected to loading in the early age, it will present a creep strain higher than for a concrete using cement with high initial strength (Mehta and Monteiro 2014).

According to Neville and Brooks (2013), the cement consumption directly affects the shrinkage once the loss of the water from the cement paste is responsible for the occurrence of such phenomenon. Therefore, higher is the cement consumption, higher is the shrinkage intensity. Also, finer is the cement, higher is the risk of cracking, because the hydration reactions will happen slower leading to a retardation on the hardening process.

Type of Aggregate

The granulometry, maximum dimension, shape and texture of the aggregate are factors which influence creep and drying shrinkage, as can be observed in Fig. 2.28 (Neville and Brooks 2013).

The maximum dimension plays a trick role, because it alone doesn't impact on the shrinkage intensity. On the other hand, larger aggregates allow the usage of concrete with low w/c ratio, leading to a lower shrinkage. One of the main important characteristics of the aggregate is its Young's modulus which is responsible for controlling the deformations, and the values of creep and shrinkage strains can increase up to 2.5 times when the concrete uses a low Young's modulus aggregate rather than a high modulus one. The other factors (shape and texture) indirectly influence creep and shrinkage, for example in the densification capacity of the concrete mixture (Neville and Brooks 2013).

Admixtures and Additives

According to Ayub et al. (2014), the mineral admixtures reduce the pore sizes in the cement paste with the increase in the pore size distribution, which reduces the porosity and permeability, and subsequently the creep and shrinkage strains. The reduction in the average pore diameter is highly and inversely dependent on the replaced cement volume and also on the age of the concrete or cement paste.

Zhang and Malhotra (1995) showed through experimental tests that the admixtures such as metakaolin and active silica can reduce the porosity and permeability of concrete, leading to a reduction on the drying shrinkage. Almost a decade later, Mazloom et al. (2004) verified that with the increase on the replacement ratio of cement by active silica leads to an equivalent decrease rate of total creep and basic creep.

The setting retarders and water reduce additives are the best in order to cause dispersion of the cement particles into water, leading to a refining in the pores of the cement paste. It's believed that those types of additives increase the basic creep and drying shrinkage strains. The main effect is indirect because the usage of additives modifies the w/c ratio, which impacts on shrinkage (Dyer 2015).

According to Brooks (1989), a special care must be taken in order to evaluate the influence of additives, because when the usage of such materials is done simultaneously (with alteration on the initial concrete recipe), the comparison between two different mixtures becomes invalid. In order to obtain a valid analysis, it would be necessary to use the additives only to enhance the mixture workability.

Relative Humidity and Temperature

The shrinkage potential varies under certain environmental conditions, as observed by Joshaghani et al. (2018) in their experimental tests. High levels of RH may limit the loss of water from the cement paste leading to a mitigation in the effects of shrinkage. Low temperatures also reduced the drying shrinkage, probably due to the slower evaporation when compared to higher temperatures. On the other hand, it was observed a contradictory behavior as the temperature was increased, the direct effect of it was the increase in the evaporation rate and also an increase in the shrinkage strain. Although, an increase in the temperature accelerates the hydration process, which can avoid the evaporation of the free water and also leads to high strength at early age, indirectly resulting in the decrease of the measured values of shrinkage strain.

As the creep is directly linked to the compressive strength, Joshaghani et al. (2018) verified the increase in the compression strength at early age for an increase in the temperature, maintaining the RH as constant. Although, for concrete specimens with 28 days which was cured at high temperature (65 °C) showed values of compressive strength lower when compared to specimens cured at lower temperatures. It may be explained by the increase in the C–S–H production at early age, which helps to develop compression strength. It was also verified that the concrete has higher compressive strength at low RH.

Element Geometry

According to Mehta and Monteiro (2014), the rate of the loss of water is controlled by the length of the path that water must go through to get expelled from concrete during creep and/or shrinkage. This phenomenon is due to the resistance to the water diffusion from inside of concrete to the environment. Therefore, the dimension and shape of a concrete element—under constant RH—determine the creep and shrinkage rate.

The dimension and shape factors can be expressed by one single value, known as theoretical or effective thickness, which is equal to the ratio between the section area and the semi-perimeter in contact with the atmosphere (Mehta and Monteiro 2014).

Age of the Concrete

The shrinkage may occur over a long period of time, although the shrinkage rate will decrease rapidly over time. Considering a concrete specimen constantly loaded for 20 years under constant environmental conditions, it's believed that 14–34% of the total shrinkage strain occurs in the first two weeks, 40–80% in three months, and 66–85% in one year. In the same way, it's believed that 25% of the total creep strain occurs in the first two weeks, 50% in three months and 75% in one year (Mehta and Monteiro 2014).

Intensity and Age of the Loading

According to Troxell et al. (1958) stated a direct proportionality between the magnitude of the constant stress applied to the structure and the measured creep strain. This affirmative was made based in the experimental data which had shown that for concrete specimens with 90 days, the measured creep strain was two times higher for a loading of 4 MPa when compared to 8 MPa. Also, lower values of creep were found for concrete specimens with 3 months when compared to specimens with 28 days.

2.2.1.4 Creep and Shrinkage Modelling

In the civil engineering practice, the time dependent average behavior of concrete is related to the physical phenomena of creep and shrinkage, and it's commonly predicted using two types of models: design code models or academic models. Examples of design code models are the *Model Code* 2010 (FIB 2012), *ACI 209* (ACI 209 2008) and *JSCE* (JSCE 2010). Examples of academic models are the *GL2000* (Gardner 2004), *B4* (Bazănt and RILEM 2015) the model proposed by Sellier et al. (2016) which will be discussed in Chap. 3.

GL2000 (GARDNER, 2004)

The model *GL2000* was firstly developed by Gardner and Lockman (2001) and then modified by Gardner (2004). This model can be used independently from the chemical additives and minerals admixtures used to produce the concrete, and also doesn't depend on the temperature and the curing. The shrinkage strain is then given by the following expression (Eq. 2.8):

$$\varepsilon_{sh}(t, t_c) = 900 \cdot k \cdot \left(\frac{30}{f_{cm28}}\right)^{\frac{1}{2}} \cdot \left(1 - 1.18h^4\right) \cdot \beta(t - t_c) \times 10^{-6} \quad (2.8)$$

where k is the shrinkage constant which depends on the type of cement; t is the age of the concrete; t_c is the age when the shrinkage begins or the end of the curing; f_{cm28} is the compression strength in 28 days; h is the relative humidity; $\beta(t - t_c)$ is a function which depends on the volume/surface ratio and the age when the shrinkage takes place (t_c).

Also, in the model *GL2000* is possible to calculate the consolidation function, which considers the creep in concrete, and it can be expressed as (Eq. 2.9):

$$J(t, t_0) = \frac{1}{E_{cmt0}} + \frac{\phi_{28}(t, t_0)}{E_{cm28}} \quad (2.9)$$

where E_{cmt0} is the Young's modulus measured at the age when the loading was applied; E_{cm28} is the Young's modulus in 28 days; $\phi_{28}(t, t_0)$ is a function which depends on the volume/ratio, RH and the age when the loading was applied (t_0).

ACI 209 (ACI, 2008)

The model *ACI 209* was developed by the American Concrete Institute (ACI 2008) and has the aim to predict the stress produced by shrinkage and creep over time, presenting the same behavior: a hyperbolic curve with an asymptotic value called as final value. The curve shape and the final value depends on the curing conditions, age when the loading was applied, mixture proportions, environmental temperature and RH, and other factors. The shrinkage strain is then given by the following expression (Eq. 2.10):

$$\varepsilon_{sh}(t, t_c) = \frac{(t - t_c)^{\alpha}}{f + (t - t_c)^{\alpha}} 780 \cdot \left(\gamma_{sh,tc} \cdot \gamma_{sh,RH} \cdot \gamma_{sh,VS} \cdot \gamma_{sh,S} \cdot \gamma_{sh,\psi} \cdot \gamma_{sh,C} \cdot \gamma_{sh,\alpha}\right) \times 10^{-6} \quad (2.10)$$

where $(t - t_c)$ is the time to initiate drying; f and α are constants which depends on the shape and size of the structural member, respectively; $\gamma_{sh,tc}$, $\gamma_{sh,RH}$, $\gamma_{sh,VS}$, $\gamma_{sh,S}$, $\gamma_{sh,\psi}$, $\gamma_{sh,C}$ and $\gamma_{sh,\alpha}$ are parameters to represent the initial curing, environmental RH, size of the structural member in terms of volume/surface ratio, slump, fine aggregates, cement content and voids percentage, respectively.

Also, in the model *ACI 209* is possible to calculate the consolidation function, which considers the creep in concrete, and it can be expressed as (Eq. 2.11):

$$J(t, t_0) = \frac{1 + \phi(t, t_0)}{E_{cmt0}} \quad (2.11)$$

where E_{cmt0} is the Young's modulus measured at the age when the loading was applied; $\phi(t, t_0)$ is the creep coefficient.

The creep coefficient may be calculated by the following expression (Eq. 2.12):

$$\phi(t, t_0) = \frac{(t - t_0)^{\psi}}{d + (t - t_0)^{\psi}} \cdot 2.35 \cdot \left(\gamma_{c,t0} \cdot \gamma_{c,RH} \cdot \gamma_{c,VS} \cdot \gamma_{c,S} \cdot \gamma_{c,\psi} \cdot \gamma_{sh,\alpha}\right) \quad (2.12)$$

where $(t - t_0)$ is the time since the loading was applied; t and d are constants to a certain shape and size of structural member; $\gamma_{c,t0}$, $\gamma_{c,RH}$, $\gamma_{c,VS}$, $\gamma_{c,S}$, $\gamma_{c,\psi}$ and $\gamma_{sh,\alpha}$ are parameters to represent the age loading factor, environmental RH, size of the structural member, slump, fine aggregates, cement content and voids percentage, respectively.

JSCE (JSCE, 2010)

The model *JSCE* developed by the Japan Society of Civil Engineers (JSCE 2010) can be found on the book *Standard specifications for concrete structures*, takes into account the properties of the aggregate and the cement, the consolidation of the concrete and its curing conditions, the shape and dimensions of the transversal section of the concrete members, the materials proportion in the mixture, and also the environmental temperature and RH. The shrinkage strain is then given by the following expression (Eq. 2.13):

$$\varepsilon_{cs}(t, t_0) = \left\{1 - \exp\exp\left[-0.108 \cdot (t - t_0)^{0.56}\right]\right\} \cdot \varepsilon_{sh} \quad (2.13)$$

where $(t - t_0)$ is the time to initiate drying; ε_{sh} is the final value of the shrinkage strain.

The final value of the shrinkage strain can be calculated as (Eq. 2.14):

$$\varepsilon_{sh} = -50 + 78 \cdot \left[1 - \exp\exp\left(\frac{RH}{100}\right)\right] + 38 \cdot \ln\ln(W) - 5 \cdot \left[\ln\ln\left(\frac{V/S}{10}\right)\right]^2 \tag{2.14}$$

where RH is the relative humidity; W is the water content in concrete; V/S is the volume/surface ratio.

Also, in the model *JSCE* is possible to calculate the creep strain by the following expression (Eq. 2.15):

$$\frac{\varepsilon_{cc}(t, t', t_0)}{\sigma_{cp}} = \left\{1 - \exp\exp\left[-0.09 \cdot (t - t')^{0.06}\right]\right\} \cdot (\varepsilon_{bc} + \varepsilon_{dc}) \tag{2.15}$$

where t, t' and t_0 represent the begin of drying, begin of loading and the loading duration, respectively; ε_{bc} and ε_{dc} represent the final values due to basic creep and drying creep, respectively; σ_{cp} is the applied stress.

The final values of the basic creep and drying creep strains can be calculated as (Eqs. 2.16 and 2.17):

$$\varepsilon_{bc} = 15 \cdot (C + W)^2 \cdot (w/c)^{2.4} \cdot \left[\ln\ln(t')\right]^{-0.67} \tag{2.16}$$

$$\varepsilon_{dc} = 4500 \cdot (C + W)^{1.4} \cdot (w/c)^{4.2} \cdot \left[\ln\ln\left(\frac{V/S}{10}\right)\right]^{-2.2} \cdot \left(\frac{1 - RH}{100}\right)^{0.36} \cdot t_0^{-0.30} \tag{2.17}$$

where C is the cement content; w/c is the water/cement ratio.

The equations shown are valid only for conventional concrete. For high-strength concrete, there are some modifications in the formulation.

Model Code 2010 (FIB, 2012)

The *Model Code 2010* was developed by the International Federation for Structural Concrete (FIB 2012) which calculates the creep and shrinkage strain with empirical equations calibrated by experimental tests. The total creep and total is divided in basic and drying creep, and the total shrinkage is divided in drying and autogenous shrinkage, regarding their different related mechanisms.

Basically, both creep strains decrease when: (i) decreasing in w/c ratio; (ii) decreasing in the cement content; (iii) increasing in the aggregate stiffness; (iv) increase of the hydration degree.

On the other hand, a decrease in the w/c ratio produces a decrease in the drying shrinkage and an increase in the autogenous shrinkage. Although a decrease in the cement content produces a decrease in both shrinkage strains.

The total creep strain is then calculated by the following expression (Eq. 2.18):

$$\varepsilon_{cc}(t, t_0) = \frac{\sigma_c(t_0)}{E_{ci}} \cdot \varphi(t, t_0) \tag{2.18}$$

where $\sigma_c(t_0)$ is the compressive stress; E_{ci} is the Young's modulus in 28 days; t is the age of the concrete; t_0 is the age when the loading starts; $\varphi(t, t_0)$ is the creep coefficient, which is the sum of the basic creep $[\varphi_{bc}(t, t_0)]$ and drying creep $[\varphi_{dc}(t, t_0)]$.

The basic and drying creep are then expressed as (Eqs. 2.19 and 2.20):

$$\varphi_{bc}(t, t_0) = \beta_{bc}(f_{cm}) \cdot \ln\ln\left(\left(\frac{30}{t_{0,adj}} + 0.035\right)(t - t_0) + 1\right) \tag{2.19}$$

$$\varphi_{dc}(t, t_0) = \beta_{dc}(f_{cm}) \cdot \beta(RH) \cdot \left(\frac{30}{0.1 + t_{0,adj}^{0.2}}\right) \ln\ln\left(\left(\frac{30}{t_{0,adj}} + 0.035\right)(t - t_0) + 1\right) \tag{2.20}$$

where $\beta_{bc}(f_{cm})$ and is the $\beta_{dc}(f_{cm})$ are the parameters related to the average compressive strength in 28 days; $\beta(RH)$ is the parameter related to the RH; $t_{0,adj}$ is the adjustment in the age of concrete related to the type of cement.

The total shrinkage strain $[\varepsilon_{cs}(t, t_s)]$ is then calculated as the sum of the drying shrinkage $[\varepsilon_{cds}(t, t_s)]$ and the autogenous shrinkage $[\varepsilon_{cas}(t)]$. Both strains can be expressed by the following expressions (Eqs. 2.21 and 2.22):

$$\varepsilon_{cds}(t, t_s) = \left[(220 + 110 \cdot \alpha_{ds1}) \cdot \exp\exp(-\alpha_{ds2} \cdot f_{cm})\right] \cdot \left[\frac{(t - t_s)}{0.035 \cdot h^2 + (t - t_\sigma)}\right] \cdot \beta(RH) \times 10^{-6} \tag{2.21}$$

$$\varepsilon_{cas}(t) = -\alpha_{as} \cdot \left[\frac{\frac{f_{cm}}{10}}{6 + \left(\frac{f_{cm}}{10}\right)}\right]^{2.5} \cdot \left[1 - \exp\left(-0.2 \cdot \sqrt{t}\right)\right] \times 10^{-6} \tag{2.22}$$

where α_{ds1}, α_{ds2} and α_{as} are the coefficients related to the type of cement; t_s is the age when drying begins; h is a coefficient linked to the geometry of the structural member.

B4 (Bazǎnt and RILEM, 2015)

The model *B4* (Bazǎnt and RILEM 2015) was developed as an enhancement of the model B3 (Bazǎnt and Baweja 2000), calibrated with a huge experimental tests database provided by the Northwestern University (Illinois, USA). The model is able to divide the total creep strain in basic and drying creep, and also is able to divide the total shrinkage strain in drying and autogenous shrinkage.

The total shrinkage strain [$\varepsilon_{sh,total}(t, t_0)$] is then calculated as the sum of the drying shrinkage [$\varepsilon_{sh}(t, t_0)$] and the autogenous shrinkage [$\varepsilon_{au}(t, t_0)$]. Both strains can be expressed by the following expressions (Eqs. 2.23 and 2.24):

$$\varepsilon_{sh}(t, t_0) = -\varepsilon_0 \cdot k_{\varepsilon a} \cdot \frac{E \cdot (7 \cdot \beta_{Th} + 600 \cdot \beta_{Ts})}{E(t_0 + \tau_{sh} \cdot \beta_{Ts})} \tag{2.23}$$

$$\varepsilon_{au}(t, t_0) = \varepsilon_{au\infty} \cdot \left[1 + \left(\frac{\tau_{au}}{t + t_0}\right)^{\alpha}\right]^{r_t}; \quad \alpha = r_a \cdot \left(\frac{w/c}{0.38}\right) \tag{2.24}$$

where $k_{\varepsilon a}$ is a factor linked to the type of aggregate; τ_{sh} is the shrinkage time interval; β_{Th} and β_{Ts} are the parameters linked to RH and temperature, respectively; ε_0 is the parameter linked to the proportion of the materials in the concrete mixture; w/c is the water/cement ratio; $\varepsilon_{au\infty}$ is the maximum autogenous shrinkage strain, calculate based on the aggregate/cement ratio and w/c ratio; τ_{au} is the autogenous shrinkage time interval; r_a and r_t are parameters linked to the type of cement.

Also, in the model *B4* is possible to calculate the consolidation function, which considers the creep in concrete, and it can be expressed as (Eq. 2.25):

$$J(t, t_0) = q_1 + C_0(t, t_0) + C_d(t, t_0, t_c) \tag{2.25}$$

where q_1 is the instantaneous strain due to unitary stress; $C_0(t, t_0)$ and $C_d(t, t_0, t_c)$ is the consolidation function for basic and drying creep, respectively; t, t_0 and t_c are the age of concrete, the age when the drying begins (curing end) and the loading age, respectively.

The two consolidation functions can be expressed as (Eqs. 2.26 and 2.27):

$$C_0(t, t_0) = q_2 \cdot Q(t, t_0) + q_3 \cdot \ln \ln\left[1 + (t - t_0)^n\right] + q_4 \cdot \ln \ln(t/t_0) \tag{2.26}$$

$$C_d(t, t_0, t_c) = q_5 \cdot \{\exp \exp[-p_{5H} \cdot H(t, t_c)] - \exp \exp[-p_{5H} \cdot H_c(t_0, t_c)]\}^{1/2} \tag{2.27}$$

where q_2, q_3, q_4 and q_5 are parameters linked to the cement content, viscoelastic consolidation, aging flux consolidation and drying creep consolidation, respectively; $Q(t, t_0)$ is the function related to the beginning of drying; $H(t, t_c)$ and $H_c(t_0, t_c)$ are the functions related to the RH, compressive strength and theoretical thickness; p_{5H} is a parameter linked to the type of cement.

2.2.2 Poro-Mechanical Theory

As previously mentioned, the swelling of the AAR gel exerts a pressure into the cement matrix. This pressure depends on the volume of gel created, on the porosity of the cement paste and also on the stiffness of each phase (fluid and solid). The behavior of the concrete over time, when subjected to a porous pressure (AAR or water), can be described applying the poro-mechanical theory.

Biot (1941) developed the poro-elasticity law (Eq. 2.28), which relates the effective stress and the total stress inside the porous media to the pressure exerted and also to the solid skeleton compressibility.

$$\overline{\overline{\sigma}} = \overline{\overline{\sigma}}' - b \cdot p \cdot \overline{\overline{1}} \tag{2.28}$$

where $\overline{\overline{\sigma}}$ is the total stress in the material; $\overline{\overline{\sigma}}'$ is the effective stress in the solid skeleton considering the porosity; p is the pressure exerted by the fluid inside the pores; b is the Biot coefficient which weights the influence of the fluid pressure over the solid skeleton in function of their correspondent stiffness.

Almost 50 years later, Coussy (1991) presented a theory to answer the continuous mechanics theory. He proposed the poro-mechanical theory with many configurations of pore-solid matrix interaction, considering a multiphase medium where the different behaviors of the fluid and the solid skeleton are taken into account. The main hypothesis of this theory are: (i) the small transformations deformations of the solid skeleton; (ii) the small variations on the porosity.

In order to describe this multiphase system, it was used the combination of balance equations, mechanical equilibrium equations (Eq. 2.29) and state equations (of the fluid, skeleton and fluid transport). The state equations were represented with Darcy's law and the balance equations were represented with a conversation law for fluid mass (Eq. 2.30).

$$div\overline{\overline{\sigma}} + \left(m_s + m_f\right) \cdot \underline{g} = 0 \tag{2.29}$$

$$\frac{\delta m_f}{\delta t} = \frac{\delta\left(\rho_f \phi\right)}{\delta t} = -\,div\,\underline{w} \tag{2.30}$$

where $\overline{\overline{\sigma}}$ is the total stress; m_s is the skeleton mass; m_f is the fluid mass, which is equal to the product of the fluid volume by the porosity $(\rho_f \phi)$; w is the water content.

The total stress can be written with the poro-elasticity theory for a linear isotropic media (Eq. 2.31):

$$\overline{\overline{\sigma}} = \lambda_0 \cdot tr\left(\overline{\overline{\varepsilon_e}}\right) \cdot \overline{\overline{1}} + 2 \cdot \mu \cdot \overline{\overline{\varepsilon}} - b \cdot p \cdot \overline{\overline{1}} \tag{2.31}$$

where $\overline{\overline{\sigma}}$ is the total stress; λ_0 and μ are the drained Lamé's coefficients (first coefficient and shear modulus); b is the Biot coefficient; p is the fluid pressure exerted on the solid skeleton.

The elastic strain (ε_e) is written as (Eq. 2.32):

$$\varepsilon_e = \varepsilon_{total} - \varepsilon_{pl} - \varepsilon_{cr} - \varepsilon_{th} \tag{2.32}$$

where ε_{total} is the total strain; ε_{pl} is the plastic strain; ε_{cr} is the creep strain; ε_{th} is the thermal strain.

According to the hypothesis of small variations on fluid mass and uniform porosity, the volume variation of fluid mass can be written as follows (Eq. 2.33):

$$v_f = b \cdot tr\left(\overline{\overline{\varepsilon_e}}\right) + \frac{p}{M} = \frac{m_f - m_{f0}}{\rho_{f0}} \tag{2.33}$$

where M is the Biot modulus.

Assuming that the pressure is homogeneously exerted by the fluid, the Biot modulus (M) characterizes the variations of fluid mass in a triaxial test with zero volumetric deformation (Eq. 2.34):

$$\frac{1}{M} = \frac{b - \varphi_0}{K_s} + \frac{\varphi_0}{K_f} \tag{2.34}$$

where φ_0 is the initial porosity accessible by the fluid; K_s is the compressive modulus of the solid; K_f is the isothermal compressive modulus of the fluid.

The effective stress (σ') may reach aa anisotropic state depending on the external stress, which can be due to structural loading in the case of concrete structures.

2.2.3 *Poro-Mechanical Models for Swelling in Concrete Structures*

In order to link the chemical models and mechanical models aforementioned, in this chapter is presented the main poro-mechanical models which includes the swelling reactions (e.g. AAR).

In order to numerically reproduce the AAR expansions, Dormieux et al. (2004) proposed a model based in a micro-poro-mechanical framework, where the concrete is divided in 3 different zones:

(i) the pores not filled with gel;
(ii) the pores filled with gel;
(iii) solid matrix (solid skeleton).

The volume of gel increases until it fills the pores, then the swelling of the solid matrix begins. Therefore, in order to use this model, it's necessary to establish the elasticity of the material and to define its microstructure (shape and volume of the pores). After, Lemarchand et al. (2005) enhanced this model through the addition of the kinetics of the gel production based on the work of Larive (1998).

Following the works of Capra and Sellier (2003), a poro-mechanical model was proposed by Grimal et al. (2018), where a phenomenological model was coupled with a damage model. The advancement of the reaction is calculated through the main environmental parameters, which are the temperature and the saturation degree (as shown in Eq. 2.7). Once the value of the advancement of the reaction (A) is determined for a given time, the intrapore pressure of gel (P_g) is then calculated as (Eq. 2.35):

$$P_g = M_g \cdot \langle A \cdot V_g - \langle A_0 \cdot V_g + b_g \cdot tr(\varepsilon) \rangle \rangle \tag{2.35}$$

where M_g is the Biot modulus which is responsible for the elastic interaction between the gel and the solid matrix; V_g is the maximum volume of gel that can be created, which is calibrated by the user; A_0 is the threshold of the advancement of the reaction, which the swelling begins; b_g is the Biot coefficient of the gel; $tr(\varepsilon)$ is the volume created by the swelling strains; $\langle\rangle$ is the Macaulay brackets.

The pressure of gel increases from a certain advancement and the state of deformation of the material influences the evolution of the pressure. The volume of the available pores to the expansion of the gel is then changed. In order to realistically reproduce the behavior of the concrete structures, this chemical model was coupled with a rheological model which takes into account creep, orthotropic damage and external stresses. The model was validated through comparison with the experimental data provided by Multon and Toutlemonde (2006), which submitted AAR-affected concrete specimens to stresses and confinement, and also by the experimental data provided by Multon et al. (2005), which analyzed concrete beams subjected to AAR, under different reinforcements distributions and water gradients.

Comi et al. (2009) proposed a poro-mechanical coupled to a chemo-thermo-damage model. The gel pressure is calculated with the advancement of the reaction which is described with the equations from Larive (1998). The anelastic effective stress (σ'') is then calculated as (Eq. 2.36):

$$\overline{\overline{\sigma''}} = \overline{\overline{\sigma}} + (\beta - b) \cdot p \cdot \overline{\overline{1}} \tag{2.36}$$

where $\overline{\overline{\sigma}}$ is the external stress; p is the gel pressure; b is the Biot coefficient; β is a coefficient to link the pressure and the damage ($\beta < b$).

Two scalars allow to take into account two isotropic damages, one for tension and other for compression. The hardening law allows producing an elastic domain in function of the damages. The model was validated through comparison with the experimental data provided by Multon and Toutlemonde (2006), and after it was applied to two concrete dams under AAR-degradation and the results were in very good agreement with the in-situ measures of cracking.

Some years later, Charpin and Ehrlacher (2014) proposed a micro-poro-mechanical model coupled with AAR orthotropic damage located in the ITZ and in the cement paste. Thus, the concrete was discretized in 4 different zones: (i) the cement paste; (ii) Interfacial transition zone (ITZ); (iii) the attacked zone of the aggregate; (iv) the unreacted aggregate.

The gel is formed in the attacked zone of the aggregate and can spread into the ITZ. The damage criteria are described by an energetic approach. Two phenomena are possible, which are a decohesion between the aggregate and the ITZ and a cement paste cracking. This model takes into account the effects of the stress in the swelling anisotropy, and it was validated through comparison with the experimental data provided by Multon and Toutlemonde (2006).

In order to enhance the model developed by Grimal et al. (2018), Salgues et al. (2014) proposed a model that could reproduce the concrete behavior under ISR (e.g. AAR and DEF). The main difference in the intrapore pressure (P_{AFt}) calculation between this model and the previous one, is the addition of the plastic strain (Eq. 2.37), calculated through an anisotropic plastic criterion (Eq. 2.38) which depends on the effective stress in the main direction.

$$P_{AFt} = M_{AFt} \cdot \langle V_{ARt} - \langle V_{0,ARt} + b_0 \cdot tr(\varepsilon) + (1 - b_0) \cdot tr(\varepsilon^p) \rangle \rangle \tag{2.37}$$

$$f_I = \left\langle \tilde{\sigma}'_I - b_0^{p^{rgi}} \right\rangle + b_1^{p^{rgi}} - b_1^{p^{lim}} \tag{2.38}$$

where M_{AFt} is the Biot modulus; V_{ARt} is the volume of the swelling product for a given time; $V_{0,ARt}$ is the pore volume to be filled by the product to initiate the swelling; b_0 is the Biot coefficient; $tr(\varepsilon)$ is the volume created by the swelling strains; $tr(\varepsilon^p)$ is the volume created by the plastic strains; p^{lim} is the maximum pressure acceptable by the unconfined concrete; b_1 is a characteristic material parameter; $\langle\rangle$ is the Macaulay brackets.

If p^{lim} is overwhelmed, $f_I > 0$ which implies that a plastic strain is propagated in one direction until the criterion becomes zero (or negative) again. Therefore, a positive hardening law allows calculating the corresponding plastic strain. The corresponding damage is then calculated with the formula (Eq. 2.39) proposed by Capra and Sellier (2003). This damage also affects on the strength and Young's modulus of the concrete.

$$\varepsilon_i^{AAR} = \varepsilon_0^{AAR} \cdot \frac{d_i^+}{1 - d_i^+} \tag{2.39}$$

where ε_i^{AAR} is the swelling strain due to AAR in one direction i; ε_0^{AAR} is the characteristic swelling strain (fixed as 0.3%); d_i^+ is the damage due to the swelling in one direction i.

Lastly, Morenon (2017) developed a poro-mechanical model to describe the swelling of AAR and DEF. This model is coupled with a rheological model proposed by Grimal et al. (2018). The full numerical implementation of these combined works can be found in Sellier (2018). The model proposed by Morenon (2017) is described in Chap. 3.

References

ACI 209 (2008) Guide for modeling and calculating shrinkage and creep in hardened concrete. American Concrete Institute (ACI), Farmington Hills

Andrade T (2006) Histórico de casos de RAA ocorridos recentemente em fundações de edifícios na região metropolitana de Recife. In: 2º Simpósio sobre Reação Álcali-Agregado em Estruturas de Concreto. Anais... IBRACON, Rio de Janeiro, pp 1–16

Andrade T, Silva JJR, Almeida R, Patrocínio J, Kihara Y, Pecchio M (2006a) Diagnóstico de reação álcali-agregado em blocos de fundação de um edifício público situado na cidade de Recife/PE. In: 2º Simpósio sobre Reação Álcali-Agregado em Estruturas de Concreto. Anais... IBRACON, Rio de Janeiro, pp 1–15

Andrade T, Silva JJR, Hasparyk N, Silva CM (2006b) Investigação do potencial de reatividade para o desenvolvimento de RAA dos agregados miúdos e graúdos comercializados na região metropolitana do Recife. In: 2º Simpósio sobre Reação Álcali-Agregado em Estruturas de Concreto. Anais... IBRACON, Rio de Janeiro, pp 1–16

Andriolo FR (2000) AAR dams affected in Brazil report on the current situation. In: Proceedings of the 11th international conference on alkali-aggregate reaction in concrete. Quebec, pp 1243–1252

Ayub T, Khan SU, Memon FA (2014) Mechanical characteristics of hardened concrete with different mineral admixtures: a review. Sci World J 2014:1–15

Balabuch TJR (2018) Análise numérica das deformações do concreto sujeito à reação álcali-agregado considerando os efeitos de retração e fluência. Master thesis, Escola de Engenharia de São Carlos, Universidade de São Paulo (USP), Brasil, p 114

Bažănt ZP, Baweja S (2000) Creep and shrinkage prediction model for analysis and design of concrete structures: model B3. In: ACI Adam Neville symposium: creep and shrinkage—structural design effects. American Concrete Institute (ACI), Farmington Hills, pp 1–83

Bažănt ZP, Panula L (1978) Practical prediction of time-dependent deformations of concrete—part II: basic creep. Matér Constr 11:317–328

Bažănt ZP, RILEM (2015) RILEM draft recommendation: TC-242-MDC multi-decade creep and shrinkage of concrete: material model and structural analysis model B4 for creep, drying shrinkage and autogenous shrinkage or normal and high-strength concretes with multi-decade applicability. Mater Struct 48:753–770

Bažănt ZP, Steffens A (2000) Mathematical model for kinetics of alkali-silica reaction in concrete. Cem Concr Res 30(3):419–428

Bažănt ZP, Zi G, Meyer C (2000) Fracture mechanics of ASR in concretes with waste glass particles of different sizes. J Eng Mech 126(3):226–232

Berra M, Mangialardi T, Paolini AE (2005) Alkali-silica reactivity criteria for concrete aggregates. Mater Struct 38:373–380

Bérubé M-A, Duchesne J, Dorion JF, Rivest M (2002) Laboratory assessment of alkali contribution by aggregates to concrete and application to concrete structures affected by alkali-silica reactivity. Cem Concr Res 32:1215–1227

Biot MA (1941) General theory of three-dimensional consolidation. J Appl Phys 12:155–164

Brito Júnior A, Ferro IP (2016) Reação Álcali-Agregado: Um breve estudo da ocorrência nos blocos da Ponte Paulo Guerra—Recife/PE. In: Congresso Brasileiro de Patologia das Construções—CBPAT. Anais... ISSN, Belém, pp 2448–1459
Brooks JJ (1989) Influence of mix proportions, platicizers and super plasticizers on creep and drying shrinkage of concrete. Mag Concr Res 41(148):145–154
Broekmans MATM (2002) The alkali-silica reaction: mineralogical and geochemical aspects of some Dutch concretes and Norwegian mylonites. Ph.D. thesis, University of Utrecht, Netherlands, p 144
Capra B (1997) Modelisation des effets mécaniques induits par les reaction alkali-granulats. Ph.D. thesis, ENS Cachan, France, p 194
Capra B, Bournazel JP (1998) Modeling of induced mechanical effects of alkali-aggregate reactions. Cem Concr Res 28(2):251–260
Capra B, Sellier A (2003) Orthotropic modelling of alkali-aggregate reaction in concrete structures: numerical simulations. Mech Mater 35(8):817–830
Carasek H, Cascudo O, Caetano G (2016) Contribuição à previsão de danos para estruturas de concreto atacadas pela reação álcali-sílica, vol 44, no 83. Concreto e Construções, Brasil, pp 30–38
Charpin L, Ehrlacher A (2014) Microporomechanics study of anisotropy of ASR under loading. Cem Concr Res 63:143–157
Collins RJ, Bareham PD (1987) Alkali-silica reaction: suppression of expansion using porous aggregate. Cem Concr Res 17(1):89–96
Comby-Peyrot I, Bernard F, Bouchard PO, Bay F, Garcia-Diaz E (2009) Development and validation of a 3D computational tool to describe concrete behavior at mesoscale. Application to the alkali-silica reaction. Comput Mater Sci 46(4):1163–1177
Comi C, Fedele R, Perego U (2009) A chemo-thermo-damage model for the analysis of concrete dams affected by alkali-silica reaction. Mech Mater 41:210–230
Coussy O (1991) Mécanique des milieu poreux. Editions Technip
Dent Glasser LS (1979) Osmotic pressure and the swelling of gels. Cem Concr Inst 9(4):515–517
Dent Glasser LS, Kataoka NNK (1981) The chemistry of alkali-aggregate reaction. Cem Concr Res 11(1):1–9
Dent Glasser LS, Kataoka NNK (1982) On the role of calcium in the alkali-aggregate reaction. Cem Concr Res 12(3):321–331
Deschenes DJ (2009) ASR/DEF-damaged bent caps: shear tests and field implications. Master of Science Engineering, The University of Texas, Austin, p 295
Diamond S (1989) ASR—another look at mechanisms. In: Proceedings of the 8th international conference on alkali-aggregate reaction, Kyoto, pp 83–94
Diamond S, Barneyback Jr RS, Struble LJ (1981) On the physics and chemistry of alkali-silica reactions. Purdue University, School of Civil Engineering, Indiana, p 10
Dormieux L, Lemarchand E, Kondo D, Fairbairn E (2004) Elements of poro-mechanics applied to concrete. Mater Struct 37:31–42
Dron R, Brivot F (1993) Thermodynamic and kinetic approach to the alkali-silica reaction—part 2: experiment. Cem Concr Res 23(1):93–103
Dunant CF, Scrivener KL (2010) Micro-mechanical modelling of alkali-silica-reaction-induced degradation using the AMIE framework. Cem Concr Res 40(4):517–525
Durand B, Roux R, Houde J, Blanchette A (1992) Free expansions and stresses in concrete related to alkali aggregate reaction. In: Proceedings of 9th ICAAR, London, pp 298–310
Dyer T (2015) A Durabilidade do Concreto. Editora Ciência Moderna, Rio de Janeiro
Esposito R, Hendriks MA (2016) A multiscale micromechanical approach to model the deteriorating impact of alkali-silica reaction on concrete. Cem Concr Compos 70:139–152
Esposito R, Hendriks MA (2019) Literature review of modelling approaches for ASR in concrete: a new perspective. Eur J Environ Civ Eng 23(11):1311–1331
FIB (2012) Model code 2010, vol 1, 65th edn. International Federation for Structural Concrete (FIB), Lausanne

Furusawa Y, Ohga H, Uomoto T (1994) An analytical study concerning prediction of concrete expansion due to alkali-silica reaction. Am Concr Inst SP 145:757–780

Gameleira CMTM, Nunes VQG, Régis PA, Oliveira RA (2016) Structural analysis of pile caps, used in bridge foundation, subject to AAR. In: Maintenance, monitoring, safety, risk and resilience of bridges and bridge networks. CRC Press, Florida, p 616

Gao XX, Multon S, Cyr M, Sellier A (2013) Alkali-silica reaction (ASR) expansion: pessimum effect versus scale effect. Cem Concr Res 44:25–33

Garcia-Diaz E, Riche J, Bulteel D, Vernet C (2006) Mechanism of damage for the alkali-silica reaction. Cem Concr Res 36(2):395–400

Gardner NJ (2004) Comparison of prediction provisions for drying shrinkage and creep of normal-strength concretes. Can J Civ Eng 31(5):767–775

Gardner NJ, Lockman MJ (2001) Design provisions for drying shrinkage and creep of normal-strength concrete. ACI Mater J 98:159–167

Grimal E (2007) Caractérisation des effets du gonflement provoqué par la réaction alcali-silice sur le comportement mécanique d'une structure en béton. Ph.D. thesis, Université Paul Sabatier Toulouse, France

Grimal E, Sellier A, Le Pape Y, Bourdarot E (2007) Creep, shrinkage, and anisotropic damage in alkali-aggregate reaction swelling mechanism—part I: a constitutive model. ACI Mater J 227–235

Grimal E, Morenon P, Sellier A, Multon S, Bourdarot E (2018) AAR and DEF structural effects modelling. In: Swelling concrete in dams and hydraulic structures: DSC 2017, 1st edn. ISTE Ltd., Wiley

Hansen WC (1944) Studies relating to the mechanism by which the alkali-aggregate reaction produces expansion in concrete. J Am Concr Inst 15:213–227

Hasparyk NP (2011) Reação álcali-agregado no concreto. In: Isaia CG (ed) Concreto: ciência e tecnologia, Cap 27. IBRACON, São Paulo, pp 933–1001

Hawlett P (2003) Lea's chemistry of cement and concrete, 4th edn. Butterworth-Heinemann, p 1092

Hobbs DW (1988) Alkali-silica reaction in concrete. Thomas Telford, London, p 183

Ichikawa T, Miura M (2007) Modified model of alkali-silica reaction. Cem Concr Res 37(9):1291–1297

Idorn GM (2001) A discussion of the paper "Mathematical model for kinetics of alkali-silica reaction in concrete" by Zdenek P. Bazănt and Alexander Steffens. Cem Concr Res 31(7):1109–1110

Jensen V (2004) Alkali-silica reaction damage to Elgeseter Bridge, Trondheim, Norway: a review of construction, research and repair up to 2003. Mater Charact 53(2–4):155–1770

Joshaghani A, Balapour M, Ramezanianpour A (2018) Effect of controlled environmental conditions on mechanical, microstructural and durability properties of cement mortar. Constr Build Mater 164:134–139

JSCE (2010) Standard specifications for concrete structures—2007. Design, vol 15. Japan Society of Civil Engineers (JSCE), Tokyo

Kuperman SC (2007) Considerações sobre fluência de concretos. Téchn Rev Tecnol Constr 125:58–63

Larive C (1997) Apports combinés de l'expérimentation et de la modélisation à la compréhension de l'alcali-réaction et de ses effets mécaniques. Ph.D. thesis, Ecole Nationale des Ponts et Chaussées, Paris, p 335

Larive C (1998) Combined contribution of experiments and modeling to the understanding of alkali-aggregate reaction and its mechanical consequences. Report OA 28. Laboratoire Central des Ponts et Chaussées, Paris

LCPC (1994) Recommandations pour la prevention des désordres dus à l'alcali-réaction

Léger P, Côté P, Tinawi R (1996) Finite element analysis of concrete swelling due to alkali-aggregate reactions in dams. Comput Struct 60:601–611

Lemarchand E, Dormieux L, Ulm F-J (2005) Micromechanics investigation of expansive reactions in chemoelastic concrete. Philos Trans R Soc A Math Phys Eng Sci 363(1836):2581

Lindgård J, Nixon PJ, Borchers I, Schouenborg B, Wigum BJ, Haugen M, Åkesson U (2010) The EU "PARTNER" Project—European standard tests to prevent alkali reactions in aggregates: final results and recommendations. Cem Concr Res 40:611–635

Lindgård J, Thomas MDA, Sellevold EJ, Pedersen B, Andiç-Çakir Ö, Justnes H, Rønning TF (2013) Alkali-silica reaction (ASR)—performance testing: influence of specimen pre-treatment, exposure conditions and prism size on alkali leaching and prism expansion. Cem Concr Res 53:68–90

Lombardi J, Massard P, Perruchot A (1997) Mesure expérimentale de la cinétique de formation d'un gel silicocalcique, produit de la réaction alcalis-silice. Cem Concr Res 27:1379–1391

Madureira EL (2007) Simulação Numérica do Comportamento Mecânico de Elementos de Concreto Armado Afetados pela Reação Álcali-Agregado. Thesis (D.Sc. in civil engineering), Universidade Federal de Pernambuco, Recife, p 219

Mazloom M, Ramezanianpour AA, Brooks JJ (2004) Effect of silica fume on mechanical properties of high-strength concrete. Cem Concr Compos 26(4):347–357

Mehta PK, Monteiro PJM (2014) Concreto: estrutura, propriedades e materiais, 2ª edn. In: Hasparyk NP, Helene P, Paulon VA (Reviewers) Translate of the 4th edition in English with Concrete: microstructure, properties and materials. IBRACON, São Paulo

Morenon P (2017) Modélisation des reactions de gonflement interne des bétons avec prise en compte des couplages poro-mécaniques et chimiques. Ph.D. thesis, Université Toulouse 3 Paul Sabatier, France, p 272

Multon S (2003) Évaluation expérimentale et théorique des effets mécaniques de l'alcali-réaction sur des structures modèles. Ph.D. thesis, Université de Marne-la-Vallée, France, p 423

Multon S, Sellier A (2016) Multi-scale analysis of alkali-silica reaction (ASR): impact of alkali leaching on scale effects affecting expansion tests. Cem Concr Res 81:122–133

Multon S, Toutlemonde F (2006) Effect of applied stresses on alkali-silica reaction induced expansions. Cem Concr Res 36:912–920

Multon S, Seignol J-F, Bourdarot E, Jeanpierre A, Toutlemonde F (2005) Effets structuraux de l'alcali-réaction: Apports d'une expérimentation sur éléments de structures à la validation de modèles. Rev Eur Gén Civ 9:1219–1247

Multon S, Cyr M, Sellier A, Leklou N, Petit L (2008) Coupled effects of aggregate size and alkali content on ASR expansion. Cem Concr Res 40:508–516

Multon S, Cyr M, Sellier A, Diederich P, Petit L (2010) Effects of aggregate size and alkali content on ASR expansion. Cem Concr Res 38:350–359

Multon S, Sellier A, Cyr M (2009) Chemo-mechanical modelling for prediction of alkali silica reaction (ASR) expansion. Cem Concr Res 39:490–500

NBR 15577-1 (2008) Agregados—Reatividade álcali-agregado—Parte 1: Guia para avaliação da reatividade potencial e medidas preventivas para uso de agregados em concreto. São Paulo

NBR 15577-4 (2008) Agregados—Reatividade álcali-agregado—Parte 4: Determinação da expansão em barras de argamassa pelo método acelerado. São Paulo

Neville AM, Brooks JJ (2013) Tecnologia do Concreto. Bookman, Porto Alegre

Nguyen VP, Stroeven M, Sluys LJ (2012) Multiscale failure modeling of concrete: micromechanical modeling, discontinuous homogenization and parallel computations. Comput Methods Appl Mech Eng 201:139–156

Nguyen M, Timothy J, Meschke G (2014) Numerical analysis of multiple ion species diffusion and alkali-silica reaction in concrete. In: Computational modellings of concrete structures (EURO-C). CRC Press, Taylor & Francis Group, St Anton am Alberg

Oberholster RE (1983) Alkali reactivity of siliceous rock aggregates: diagnosis of the reaction, testing of cement and aggregate and prescription of preventive measures. In: Proceedings of the 6th international conference alkalis in concrete—research and practice, Copenhagen

Pacelli WA (1999) Tema II: Casos Históricos—Barragens e outras estruturas. In: Relatório Final—1º Simpósio sobre Reatividade Álcali-Agregado em Estruturas de Concreto. Annals... CBGB/FURNAS, Goiânia

Pan JW, Feng YT, Wang JT, Sun QC, Zhang CH, Owen DRJ (2012) Modeling of alkali-silica reaction in concrete: a review. Struct Civ Eng 6:1–18

Pecchio M, Kihara Y, Battagin AF, Andrade T (2006) Produtos da reação álcali-silicato em concretos de edificações da região metropolitana de Recife. In: 2º Simpósio sobre Reação Álcali-Agregado em Estruturas de Concreto. Anais... IBRACON, Rio de Janeiro, pp 1–16

Pignatelli R (2012) Modeling of degradation induced by alkali-silica reaction in concrete structures. Ph.D. thesis, Politecnico di Milano, Italy, p 163

Pignatelli R, Comi C, Monteiro P (2013) A coupled mechanical and chemical damage model for concrete affected by alkali-silica reaction. Cem Concr Res 53:196–210

Ponce J, Batic O (2006) Different manifestations of the alkali-silica reaction in concrete according to the reaction kinetics of the reactive aggregate. Cem Concr Res 36:1148–1156

Poole AB (1992) Alkali-silica reactivity mechanisms of gel formation and expansion. In: Proceedings of the 9th international conference on alkali-aggregate reaction, London, pp 782–789

Pourbehi MS, Van Zijl GPAG, Strasheim JAvB (2018) Modelling of alkali silica reaction in concrete structures for rehabilitation intervention. In: 5th international conference on concrete repair, rehabilitation and retrofitting (ICCRRR), Cape Town, South Africa. MATEC Web Conf 199:03007

Powers TC, Steinour HH (1955a) An interpretation of some published researchers on alkali-aggregate reaction: part 1—the chemical reactions and mechanism of expansion. Am Concr Inst J Proc 51:497–516

Powers TC, Steinour HH (1955b) An interpretation of some published researchers on alkali-aggregate reaction: part 2—a hypothesis concerning safe and unsafe reactions with reactive silica in concrete. Am Concr Inst J Proc 51:785–812

Poyet S (2003) Etude de la degradation des ouvrages en beton atteints par la reaction alkali-silice—approche experimentale et modelisation numerique des degradations dans un environnement hydro-chemo-mechanique variable. Ph.D. thesis, Université de Marne la Vallée, France, p 237

Poyet S, Sellier A, Capra B, Thèvenin-Foray G, Torrenti J-M, Tournier-Cognon H, Bourdarot E (2006) Influence of water on alkali-silica reaction: experimental study and numerical simulations. J Mater Civ Eng 18:588–596

Rüsch H, Jungwirth D, Hilsdorf HK (1983) Creep and shrinkage—their effect on the behavior of concrete structures, 1st edn. Springer-Verlag, New York, p 284

Salgues M, Sellier A, Multon S, Bourdarot E, Grimal E (2014) DEF modelling based on thermodynamic equilibria and ionic transfers for structural analysis. Eur J Environ Civ Eng 18(4):377–402

Sanchez L, Multon S, Sellier A, Cyr M, Fournier B, Jolin M (2014) Comparative study of a chemo-mechanical modelling for alkali silica reaction (ASR) with experimental evidences. Constr Build Mater 72:301–315

Sanchez LFM, Fournier B, Jolin M, Duchesne J (2015) Reliable quantification of AAR damage through assessment of the damage rating index (DRI). Cem Concr Res 67:74–92

Saouma VE, Martin RA, Hariri-Ardebili MA, Katayama TA (2015) A mathematical model for the kinetics of the alkali-silica chemical reaction. Cem Concr Res 68:184–195

Sellier A (2018) Anisotropic damage and visco-elasto-plasticity applied to multiphasic materials. Research report. LMDC—Laboratoire Matériaux et Durabilité des Constructions de Toulouse; Université de Toulouse III—Paul Sebatier, INSA de Toulouse

Sellier A, Bournazel JP, Mébarki A (1995) Modeling the alkali-aggregate reaction with descriptions of the local destructive phenomena involved. Mater Struct 28:373–383

Sellier A, Multon S, Buffo-Lacarrière L, Vidal T, Bourbon X, Guillaume C (2016) Concrete creep modelling for structural applications: non-linearity, multi-axiality, hydration, temperature and drying effects. Cem Concr Res 79:301–315

Silva NG, Collodetti G, Pichetti DZCM, Gleize PJP (2009) Efeitos do ar incorporado nas propriedades do estado endurecido em argamassas de cimento e areia. In: Anais do 51° Congresso Brasileiro do Concreto (CBC)

Silveira JFA (1997) A Expansão do Concreto em Barragens Afetadas pela RAA e a Importância das Tensões Confinantes. In: 1° Simpósio sobre Reatividade Álcali-Agregado em Estruturas de Concreto. Annals... CBGB/FURNAS, Goiânia

Stanton TE (1940) Expansion of concrete through reaction between cement and aggregates. Proc ASCE N Y 66(10):1781–1811

Suwito A, Jin W, Xi Y, Meyer C (2002) A mathematical model for the pessimum effect of ASR in concrete. Concr Sci Eng RILEM 4:23–32

Thomas MDA, Fournier B, Folliard KJ (2013) Alkali-aggregate reactivity (AAR) facts book. Report No. FHWA-HIF-13-019. U.S. Department of Transportation—Federal Highway Administration, Springfield, VA, p 211

Tiecher F (2006) Reação álcali-agregado: avaliação do comportamento de agregados do sul do Brasil quando se altera o cimento utilizado. Master thesis, Universidade Federal do Rio Grande do Sul, Porto Alegre, p 182

Troxell GE, Raphael JM, Davis RE (1958) Long-time creep and shrinkage tests of plain and reinforced concrete. Proc ASTM Int 58:1101–1120

Ulm F-J, Coussy O, Larive C, Kefei L (2000) Thermo-chemo-mechanics of ASR expansion in concrete structures. J Eng Mech 126:233–242

Wawrzeczyk J, Molendowska A (2017) Impact of air entraining method on the resistance of concrete to internal cracking. IOP Conf Ser Mater Sci Eng 245

Wendner R, Hubler MH, Bažănt ZP (2015) Statistical justification of model B4 for multidecade concrete creep using laboratory and bridge databases and comparisons to other models. Mater Struct 48:815–833

Wood JG, Young JS, Ward DE (1987) The structural effects of alkali-aggregate reaction on reinforced concrete. In: Graham-Bellew PE (ed) Concrete alkali-aggregate reactions, proceeding of the 7th international conference. Noyes Publications, Park Ridge, NJ, Ottawa

Wu T, Temizer I, Wriggers P (2014) Multiscale hydro-thermo-chemo-mechanical coupling: application to alkali-silica reaction. Comput Mater Sci 84:381–395

Zhang MH, Malhotra VM (1995) Characteristics of a thermally activated alumino-silicate pozzolanic material and its use in concrete. Cem Concr Res 25(8):1713–1725

Zhang C, Wang AQ, Tang MS, Wu BQ, Zhang NS (1999) Influence of aggregate size and aggregate size grading on ASR expansion. Cem Concr Res 29(9):1393–1396

Chapter 3
Materials and Methods

In this chapter is presented an explanation of the models used to analyze the chemo-mechanical behavior on AAR-affected concrete. The first part is a presentation of the rheological scheme (Grimal et al. 2017) that couples all the following models presented in this chapter. The second part is dedicated to describe the chemical model proposed by Morenon (2017) which shows how the advancement of the reaction creates a volume of gel, and such volume exerts an intrapore pressure in the solid skeleton. However, in order to calculate this gel pressure, the user needs the creep and shrinkage strains, and also the diffuse cracking produced by the gel (micro-plastic strains).

Therefore, the third part is dedicated to describe the creep and shrinkage models proposed by Sellier et al. (2016), in order to obtain the associated strains. The fourth and last part is dedicated to how those implementations were made in the software COMSOL Multiphysics®, showing the main features used in the software to add the models aforementioned.

3.1 Rheological Scheme (Grimal et al. 2017)

As previously mentioned, the concrete behavior is modelled in the poro-mechanical framework. The rheological scheme (Fig. 3.1) couples the damage model (Sellier et al. 2013), chemical model (Morenon 2017) and the creep and shrinkage models (Sellier et al. 2016), in order to consider the realistic interaction between ASR, creep, shrinkage and damage.

In Fig. 3.1, P_g is the pressure induced by AAR gel, P_w is the equivalent capillary pressure causing shrinkage in non-saturated conditions, σ_{ij} is the total stress, $\tilde{\sigma}_{ij}$ is the effective stress in the sense of damage and $\tilde{\sigma}'_{ij}$ is the effective stress in the sense of poro-mechanics.

F. A. N. Silva et al., *Advanced Chemical and Creep Modeling for Alkali-Aggregate Reaction in Concrete*, Building Pathology and Rehabilitation 28,
https://doi.org/10.1007/978-3-031-53980-0_3

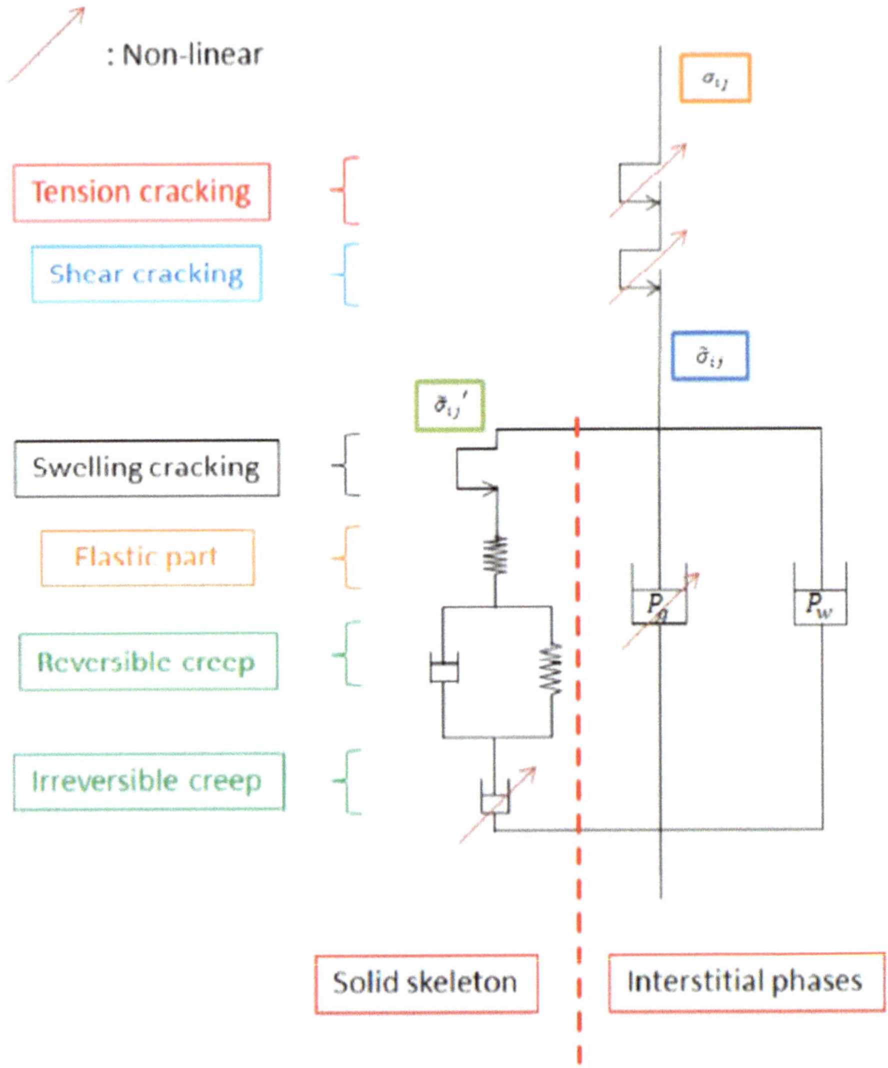

Fig. 3.1 Rheological scheme (adapted from Grimal et al. 2017)

The interstitial pressures (P_g and P_w) are taken into account with a Biot coefficient for both ASR (b_{asr}) and capillary pressure (b_w). The effective stress in the sense of damage ($\tilde{\sigma}_{ij}$) is then expressed as (Eq. 3.1):

$$\tilde{\sigma}_{ij} = \tilde{\sigma}'_{ij} - b_{asr} \cdot P_{asr} - b_w \cdot P_w \tag{3.1}$$

The effective stress increment in the sense of poro-mechanics is then calculated as (Eq. 3.2):

$$\dot{\tilde{\sigma}}_{ij} = S_0 \cdot \left(\dot{\varepsilon} - \dot{\varepsilon}_{pl} - \dot{\varepsilon}_{cr} - \dot{\varepsilon}_{th}\right) \tag{3.2}$$

where S_0 is the stiffness matrix; $\dot{\varepsilon}$, $\dot{\varepsilon}_{pl}$, $\dot{\varepsilon}_{cr}$ and $\dot{\varepsilon}_{th}$ are the total strain increment, plastic strain increment, creep strain increment and thermal strain increment, respectively.

The total stress depends on the effective stress in the sense of damage ($\tilde{\sigma}_{ij}$) and on the two scalar damage variables regarding the damage caused by ASR (diffuse cracking—D_{ASR}) and the damage caused by structural loading (D_{struct}). Therefore, the total stress is written as (Eq. 3.3):

$$\sigma_{ij} = (1 - D_{ASR}) \cdot (1 - D_{struct}) \cdot \tilde{\sigma}_{ij} \tag{3.3}$$

3.2 Chemical Model (Morenon 2017)

The AAR is responsible for the appearance of newly formed phases in the porous matrix of the concrete. The lack of space at their disposal quickly puts the solid skeleton under pressure, which results in tensile stresses around the reactive sites (aggregates). Therefore, the theory of poro-mechanics allows to take into account the influence of this intrapore pressure on the solid skeleton of the concrete (Morenon 2017). The physical phenomena responsible for the degradation of the mechanical behavior of ASR-affected concrete structures can be listed as follows:

1. **Creation of the ASR gel**: A volume of intrapore swelling product (ASR gel) is created by chemical reactions;
2. **Creation of the intrapore pressure**: When all the porosity volume around the reactive site (silica aggregate) is filled with the reaction product, a intrapore pressure is generated;
3. **Diffuse cracking appearance**: The intrapore pressure produces a tensile stress in the cement paste, and basically, when the tensile strength is reached, diffuse cracks appear;
4. **Creep induced by the intrapore pressure**: As the gel pressure is maintained over time on the solid skeleton, creep will be generated;
5. **Localized cracks appearance**: When the concrete structure is subjected to external loading, localized cracks may be generated, which will facilitate the penetration of water and other pathologies agents (e.g. CO_2), and such water may accelerate or reactivate the swelling reaction (ASR).

According to Morenon (2017), the **advancement of the reaction** (A^{asr}) is based on the works of Poyet et al. (2006) and Grimal et al. (2007). The term A^{asr} ranges from 0 (before the reaction begins) to 1 (when the reaction is complete). The evolution of the of the advancement is written as follows (Eq. 3.4):

$$\frac{\delta A^{asr}}{\delta t} = \frac{1}{\tau_{ref}^{asr}} \cdot C^{T,asr} \cdot C^{W,asr} \cdot \left\langle Sr - A^{asr}\right\rangle^{+} \tag{3.4}$$

where τ_{ref}^{asr} is the characteristic time (calibrated using a free swelling test); Sr is the saturation degree; $C^{T,asr}$ is the coefficient for temperature effect; $C^{W,asr}$ is the coefficient for humidity effect.

The **effect of temperature** ($C^{T,asr}$) is formulated with an Arrhenius' law (Eq. 3.5), as proposed by Ulm et al. (2000).

$$C^{T,asr} = \exp\left[-\frac{E^{asr}}{R} \cdot \left(\frac{1}{T} - \frac{1}{T_{ref}}\right)\right] \tag{3.5}$$

where E^{asr} is the thermal activation energy (40.000 J/mol—Larive 1997); T_{ref} is the reference temperature (calibrated on the same test as τ_{ref}^{asr}); R is the perfect gas constant [8.1314 J/(mol K)]; T is the temperature (K).

The **effect of humidity** ($C^{W,asr}$) is evaluated by a potential law (Eq. 3.6), which can describe the non-linear increase in the impact of humidity in ASR, especially the role of the saturation degree.

$$C^{W,asr} = \begin{cases} \left(\frac{Sr - Sr^{th,asr}}{1 - Sr^{th,asr}}\right)^2 & if\ Sr > Sr^{th,asr} \\ 0 & if\ Sr \leq Sr^{th,asr} \end{cases} \tag{3.6}$$

where $Sr^{th,asr}$ is the saturation degree threshold which is the minimum saturation degree to initiate the chemical reaction.

As a result of the chemical reactions, a **volume of ASR gel** (ϕ^{asr}) is created for a given time "t" (Eq. 3.7):

$$\phi^{asr}(t) = \phi^{asr,\infty} \cdot A^{asr}(t) \tag{3.7}$$

where $\phi^{asr,\infty}$ is the maximum volume ratio of gel that can be created, which is calibrated by the free swelling test (for the same condition as τ_{ref}^{asr}).

Next, on Table 3.1, there is a resume of the main material parameters used to calculate the volume of ASR gel, as well as the values range, its units and the calibration tests required.

The volume of ASR gel produced is used to calculate **the intrapore pressure** (P_{asr}) induced by the swelling of the gel (Eq. 3.8). For each instant, the pressure is calculated by the difference between the volume of ASR gel produced (ϕ^{asr}) and the available volume for the gel to permeate (ϕ_{asr}^{v}).

$$P_{asr} = M_{asr} \cdot \left\langle \phi^{asr} - \left\langle \phi_{asr}^{v} \cdot \left(\frac{P_{asr}}{\tilde{R}_I^t}\right) + b_{asr} \cdot tr\left(\varepsilon^e + \varepsilon^{cr}\right) + tr\left(\varepsilon^{p,asr}\right) \right\rangle^+ \right\rangle^+ \tag{3.8}$$

where M_{asr} is the Biot modulus, which is responsible for the interaction between ASR gel and cement matrix; b_{asr} is the Biot coefficient used to take into account the increase in porosity volume due to the evolution of the intrapore pressure; ϕ_{asr}^{v} is the

Table 3.1 Material parameters used to calculate the volume of gel (Morenon 2017)

Parameter	Signification	Range value in laboratory tests	Unit	Calibration test
$\phi^{asr,\infty}$	Maximum volume ratio of ASR gel	$0 \sim 10^{-2}$	m^3/m^3	Free swelling test
τ_{ref}^{asr}	Characteristic time of ASR	0 ~ 200	days	Free swelling test
$Sr^{th,asr}$	Saturation degree threshold to activate ASR	0 ~ 0.1	–	Swelling test for different hydric conditions
T_{ref}	Reference temperature at which the test is carried out	Temperature value	°C	–
E^{asr}	Activation energy for the kinetics of ASR	40.000 (Larive 1997)	J/mol	Swelling test for different temperature

available voids volume; $\tilde{R}_I^t$ is the micro-tensile strength of the material; $tr(\varepsilon^e + \varepsilon^{cr})$ is variation of the porosity due to concrete strain, and its composed by the trace of the elastic strains (ε^e) and creep strain (ε^{cr}) of the cement matrix; $tr(\varepsilon^{p,asr})$ is the variation of the porosity due to plastic strain, representing the volume of cracks produced by ASR and assumed to be completely filled by ASR gel, and its composed by the trace of the plastic strains ($\varepsilon^{p,asr}$) of the cement matrix; $\langle\rangle^+$ is the Macaulay bracket, which denotes the positive part operator.

The available volume for the permeation of the gel (φ_{asr}^v) is influenced by many factors, such as:

- Connected porosity;
- Increase or decrease in the volume of voids where the gel is formed (due to elastic strains or creep of the cement matrix);
- Volume of the cracks created by the ASR.

Next, on Table 3.2, there is a resume of the main material parameters used to calculate the volume of ASR gel, as well as the values range, its units and the calibration tests required.

Table 3.2 Material parameters to calculate the intrapore pressure (Morenon 2017)

Parameter	Signification	Range value in laboratory tests	Unit	Calibration test
ϕ_{asr}^v	Volume of porosity connected to the reactive sites	$\approx 10\%\ of\ \phi^{asr}$	m^3/m^3	Free swelling test
b_{asr}	Biot coefficient for ASR	0.1 ~ 0.25	–	Restrained swelling test
M_{asr}	Biot modulus for gel-matrix interaction	27.700	MPa	Free swelling test

3.3 Creep and Shrinkage Models (Sellier et al. 2016)

The poro-mechanical creep and shrinkage models can be summarized in a rheological scheme (see Fig. 3.2) which contains two branches:

- **Left branch—Solid skeleton**: the elastic part (symbolized as a spring) is used to model the instantaneous behavior. The Kelvin module (symbolized as spring and a Newtonian dashpot in parallel) is used to model the reversible creep. The non-linear Maxwell module (symbolized as a Newtonian dashpot) is used to model the permanent strains.
- **Right branch—Interstitial phases**: represents the effects of hydric forces (capillary pressure and variation of disjoining forces) and also the pressure generated by swelling products (e.g., AAR gel).

Although the concrete microstructure is not explicitly modelled, the constitutive equations reflect the several underlying phenomena induced by concrete heterogeneity. The viscous phenomenon is assumed to happen in the C–S–H interlayers, while the other hydrates and aggregates are considered as non-viscous elements.

3.3.1 *Permanent Creep—Maxwell Strain*

The permanent creep is modeled with a Maxwell module (Fig. 3.3) and it's assumed to derive directly from the corresponding elastic strain components (ε_I^E). The proportionality between the creep velocity and the ε_I^E is supported by experimental data which shows that multiaxial creep starts with relative strain velocities in different directions (I) proportional to the ε_I^E. Therefore, the Maxwell strain component (ε_I^M) is written as (Eq. 3.9):

$$\frac{\partial \varepsilon_I^M}{\partial t} = \frac{\varepsilon_I^E}{\tau_I^M} \tag{3.9}$$

where τ_I^M is the characteristic time associated with the ε_I^M; I is the eigenvalue to represent the main directions.

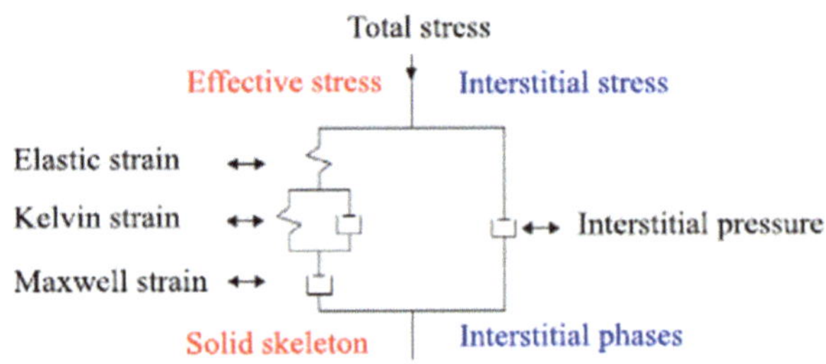

Fig. 3.2 Idealized rheological scheme for poro-mechanical creep model (adapted from Sellier et al. 2016)

Parameters

Label: Parameters - General

▾ Parameters

Name	Expression	Value	Description
Sr	0.99	0.99	Saturation Degree
T	273.15+25 [K]	298.15 K	Temperature
R	8.1314 [J/(mol*K)]	8.1314 J/(mol·K)	Perfect Gas Constant
Fy	5 [MPa]	5E6 Pa	Pressure applied on test
Delta	0.1 [d]	8640 s	Timestep
tf	Delta*500	4.32E6 s	Total time
tk	Delta*2500	2.16E7 s	Time when start unloading

Fig. 3.3 General parameters—environmental conditions, loading and total time of simulation

The characteristic time is then expressed as a function (Eq. 3.10) of the reference characteristic time (τ_{ref}^{M}), characterizing the initial material state, and also the consolidation function (C_I^C). The τ_{ref}^{M} is a fitting parameter which is inversely proportional to the initial creep velocity, under controlled conditions (temperature, humidity and mechanical loading), and the C_I^C considers the non-linear dependence of the creep potential on the loading rate.

$$\tau_I^M = \tau_{ref}^M \cdot C_I^C \tag{3.10}$$

As the consolidation function is written as an anisotropic formulation (Eq. 3.11), it considers the different consolidation velocities in the three main directions of creep. The objective of the consolidation function is to consider the decrease of the creep strain velocity with time without relying on the time variable itself.

$$C_I^C = \frac{1}{k} \cdot \exp\left(\frac{1}{k} \cdot \left\langle \frac{\varepsilon_I^M}{\varepsilon_I^E} \right\rangle^+ \right) \tag{3.11}$$

where k is the creep coefficient for the same conditions as τ_{ref}^M; $\langle\rangle^+$ is the Macaulay bracket which denotes the positive part operator.

The creep coefficient is affected by physical conditions such as temperature (T), humidity (H) and mechanical loading (M), thus it can be expressed as (Eq. 3.12):

$$k = k_{ref} \cdot C^T \cdot C^H \cdot C^M \tag{3.12}$$

where k_{ref} is the reference creep coefficient; C^T, C^H and C^M are the functions related to the influence of temperature, humidity and mechanical loading, respectively.

The reference creep coefficient is then written as (Eq. 3.13):

$$k_{ref} = \frac{\varepsilon_{ref}^M}{\varepsilon_{ref}^E} \tag{3.13}$$

where ε_{ref}^{E} is the reference elastic strain which is a fitting parameter obtained for a given loading level applied on experimental test; ε_{ref}^{M} is the reference creep potential which is a fitting parameter related to the creep function amplitude, and it's defined for the same conditions as ε_{ref}^{E}.

The reference elastic strain is the immediate strain related to the creep test for a loading of 30% (or less) of its compressive strength. In the case of real structures which doesn't have creep tests performed, this strain is often estimated as a ration between the uniaxial compressive strength (R_c) and the Young's modulus (E), expressed as follows (Eq. 3.14):

$$\varepsilon_{ref}^{E} = \frac{\left(\frac{R_c}{3}\right)}{E} \tag{3.14}$$

The function related to the influence of humidity (C^H) is taken as equal to the saturation degree (S_r), written as (Eq. 3.15):

$$C^H = S_r \tag{3.15}$$

The function related to the influence of the temperature (C^T) is taken as a product of two functions (Eq. 3.16): the first is related to the influence of temperature in water viscosity (C_w^T) and the second is related to the influence of temperature in the differential dilatation between non-viscous and viscous phases (C_p^T).

$$C^T = C_w^T \cdot C_p^T \tag{3.16}$$

The two functions C_w^T and C_p^T are then written as Arrhenius laws as follows (Eqs. 3.17 and 3.18):

$$C_w^T = \exp\left[-\frac{E_w^a}{R} \cdot \left(\frac{1}{T} - \frac{1}{T_{ref,creep}}\right)\right] \tag{3.17}$$

$$C_p^T = \begin{cases} \exp\left[-\frac{E_p^a}{R} \cdot \left(\frac{1}{T} - \frac{1}{T_{thr}}\right)\right] & if\ T > T_{thr} \\ 1 & if\ T \leq T_{thr} \end{cases} \tag{3.18}$$

where E_w^a is the activation energy of water viscosity (≈ 17,000 J/mol); R is the perfect gas constant [8.1314 J/(mol K)]; T is the environment temperature; $T_{ref,creep}$ is the reference temperature for which the τ_{ref}^{M} is fitted; E_p^a is the activation energy for differential dilatation between phases (≈ 25,000 J/mol); T_{thr} is the temperature threshold (≈ 45 °C) for which the thermal damage appears and modifies the creep potential.

The function related to the influence of the mechanical loading (C^M) starts from 1 (for weakly loaded material) and diverges if the loading level reaches a critical value causing tertiary creep, which is believed to occur only in cases of micro-structure damage. Therefore, C^M is associated an equivalent Drucker–Prager criterion, and it's written as (Eq. 3.19):

$$C^M = \frac{\tau_{cr}^{DP}}{\tau_{cr}^{DP} - \tau^{DP}}; \quad \tau_{cr}^{DP} > \tau^{DP} \tag{3.19}$$

where τ_{cr}^{DP} is the critical stress leading to tertiary creep; τ^{DP} is the Drucker–Prager equivalent shear stress.

The two stresses (τ^{DP} and τ_{cr}^{DP}) are then expressed as follows (Eqs. 3.20 and 3.21):

$$\tau^{DP} = \sqrt{\frac{\sigma^D : \sigma^D}{2}} + \delta \cdot \frac{tr(\sigma)}{3} \tag{3.20}$$

$$\tau_{cr}^{DP} = \frac{\sigma_{cr}}{\sqrt{3}} \cdot \left(1 - \frac{\delta}{\sqrt{3}}\right) \tag{3.21}$$

where σ^D is the deviatoric part of the stress's tensor; $tr(\sigma)/3$ is the hydrostatic pressure δ is the Drucker–Prager confinement coefficient; σ_{cr} is the uniaxial critical stress assumed as a compression.

The uniaxial critical stress is then written as (Eq. 3.22):

$$\sigma_{cr} = \frac{2}{3} \cdot \left(\frac{\chi^M}{\chi^M - 1}\right) \cdot R_c \tag{3.22}$$

where R_c is the uniaxial compressive strength; χ^M which is a fitting parameter responsible for the non-linear behavior when the specific creep amplification increases with the loading rate.

Next, on Table 3.3, there is a resume of the main material parameters used to calculate the reversible creep, as well as the values range and its units.

3.3.2 *Reversible Creep—Kelvin Strain*

The reversible creep is modeled with a Kelvin module (Fig. 3.3) and the asymptotic strain is assumed to be proportional to the elastic strain components (ε_{ij}^E). The directions ij correspond to the base in which the stress tensor is expressed. Therefore, the Kelvin strain component (ε_{ij}^K) is written as (Eq. 3.23):

$$\frac{\partial \varepsilon_{ij}^K}{\partial t} = \frac{1}{\tau^K} \cdot \left(\frac{\varepsilon_{ij}^E}{\psi^K} - \varepsilon_{ij}^K\right) \tag{3.23}$$

Table 3.3 Material parameters to calculate the permanent creep (Sellier et al. 2016)

Parameter	Signification	Range value in creep tests	Unit
τ^M_{ref}	Reference characteristic time associated with the Maxwell strain	$\approx$ 14	days
E^a_w	Activation energy for water viscosity	17.000	J/mol
E^a_p	Activation energy for differential dilatation between phases	25.000	J/mol
$T_{ref,creep}$	Reference temperature for which the τ^M_{ref} is fitted	20	°C
T_{thr}	Temperature threshold for which the thermal damage appears and modifies the creep potential	40	°C
χ^M	Fitting parameter responsible for the non-linear behavior when the specific creep amplification increases with the loading rate	2	–
ε^M_{ref}	Reference creep coefficient	0.9×10^{-4}	–

where τ^K is the characteristic time associated with the ε^K_{ij}; ψ^K is a fitting parameter to control the kinetics modification, adjusting the final amplitude of the reversible creep, and it's obtained through creep test with strain recovery.

The characteristic time associated with the reversible creep has its velocity associated depends on the environmental condition, such as temperature—in the case of water viscosity (C^T_w)—and humidity (C^H). Therefore, it can be expressed as (Eq. 3.24):

$$\tau^K = \tau^K_{ref} \cdot C^T_w \cdot C^H \tag{3.24}$$

where τ^K_{ref} is the reference characteristic time associated with the ε^K_{ij}, calibrated for the same condition as ψ^K.

The authors (Sellier et al. 2016) have shown through experimental test database that the value of the reversible creep (Kelvin strain) represents about 20–25% of the elastic strain.

Next, on Table 3.4, there is a resume of the main material parameters used to calculate the permanent creep, as well as the values range and its units.

3.3.3 Shrinkage

The effects of intrapore water is modeled in a poro-mechanical framework because the water molecules presented in nanoscopic interlayers cause attractive or repulsive forces according to their environmental conditions (temperature, humidity and

Table 3.4 Material parameters to calculate the reversible creep (Sellier et al. 2016)

Parameter	Signification	Range value in creep tests	Unit
ψ^K	Fitting parameter to control the kinetics modification, adjusting the final amplitude of the reversible creep	4 ~ 5	–
τ^K_{ref}	Reference characteristic time associated with the Kelvin strain	0.70	days

mechanical loading), impacting on the macroscopic scale, as already explained in Chap. 2. The hydric forces effects result in stress field heterogeneity, and subsequently damage induced by this stress gradient.

For the sake of simplicity, the different effects of water forces—capillary and disjoining forces—were combined into a single **equivalent capillary pressure** (p_w) which depends on the stress state (as illustrated in Fig. 3.4). As the capillary pressures will act when the conditions are unsaturated, a Van-Genuchten law was used to describe the relation between p_w and the saturation degree (Eq. 3.25):

$$p_w = -M_{shr} \cdot \left(S_r^{\left(\frac{-1}{m_{vg}}\right)} - 1 \right)^{(1-m_{vg})} \tag{3.25}$$

where M_{shr} and m_{vg} are fitting parameters calibrated from the work of Baroghel-Bouny et al. (1999) which controls the shape of the water retention curve, and their values are 41 MPa and 0.5, respectively.

Morenon (2017) re-wrote the previous Van-Genuchten law as follows (Eq. 3.26):

$$p_w = a \cdot \left(1 - S_r^{\left(1-\frac{1}{b}\right)}\right)^{(1-b)} \tag{3.26}$$

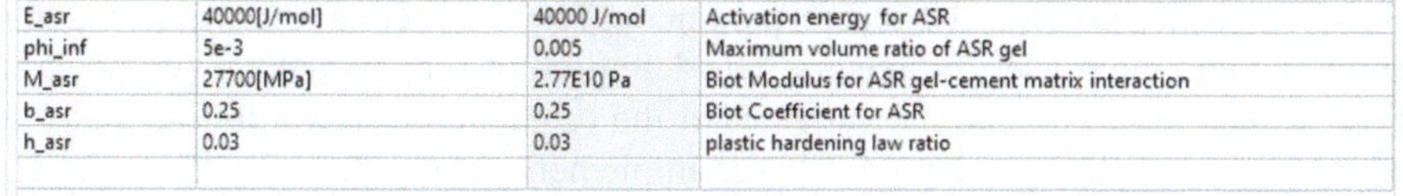
Parameters

Label: Parameters - Chemical model

Parameters

Name	Expression	Value	Description
tau_ref	50*24*3600[s]	4.32E6 s	Characteristic time of ASR
Sr_th	0.1	0.1	Saturation degree threshold to activate ASR
T_ref	38+273.15[K]	311.15 K	Reference temperature for Free Swelling Test calibration
E_asr	40000[J/mol]	40000 J/mol	Activation energy for ASR
phi_inf	5e-3	0.005	Maximum volume ratio of ASR gel
M_asr	27700[MPa]	2.77E10 Pa	Biot Modulus for ASR gel-cement matrix interaction
b_asr	0.25	0.25	Biot Coefficient for ASR
h_asr	0.03	0.03	plastic hardening law ratio

Fig. 3.4 Chemical model parameters

where a and b are the fitting parameters calibrated from the work of Multon and Toutlemonde (2006), and their values are 25 MPa and 0.42, respectively.

3.4 Implementation in COMSOL®

The models previously shown in this chapter were implemented in COMSOL Multiphysics® v. 5.5 (COMSOL AB, Stockholm, Sweden) a commercial multiphysics simulation platform. Firstly, the geometries of the concrete specimens were constructed by the graphic interface. The mechanical and chemical were realized by different models in *COMSOL*. These models were implemented through equation based modelling and they were transferred to different ordinary differential equations (ODEs) and combined into the solid mechanics interface, which can be solved using the nonlinear solver provided by *COMSOL*.

The Solid mechanics interface in *COMSOL Multiphysics®* is based on the standard momentum balance equation (Eq. 3.27), and for simplicity, inertial effects were neglected and displacements and strains were assumed to remain small (Eq. 3.28). Therefore, these two equations are complemented by appropriate boundary conditions and a constitutive law relating stresses and strains to complete the problem definition.

$$\nabla \cdot \sigma + F_V = 0 \tag{3.27}$$

$$\varepsilon = \frac{1}{2} \cdot \left(\nabla u + \nabla u^T\right) \tag{3.28}$$

where σ is the stress tensor; F_V contains the body forces; ε is the strain tensor; u are the displacements.

The results were analyzed using the post-processing functions provided by the software and verified from decoupling tests.

3.4.1 Implementation of the Chemical Model

First, it is needed to define the general parameters (environmental parameters, loading and time) and the model parameters. The definition of such parameters is shown in Figs. 3.3 and 3.4. The internal variables of the problem are shown in Fig. 3.5. The advancement of the reaction (A^{asr}) was implemented as a domain ODE, and it's shown in Fig. 3.6.

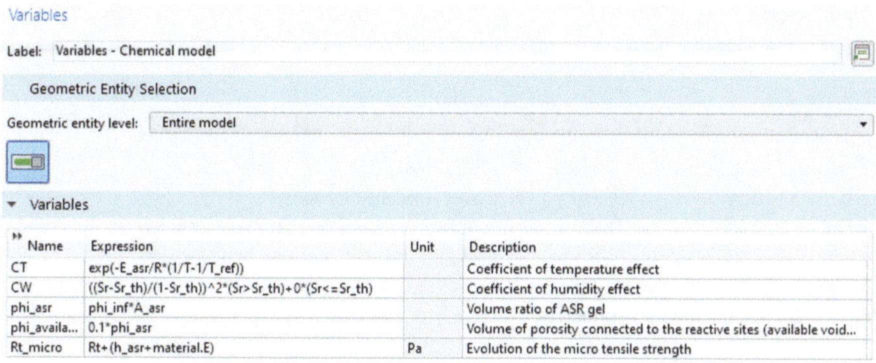

Fig. 3.5 Internal variables of the chemical model

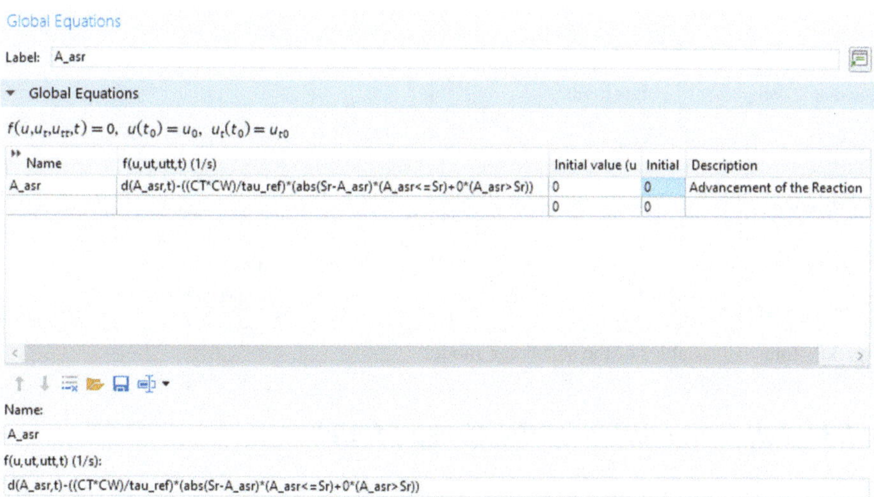

Fig. 3.6 Advancement of the reaction written as a domain ODE

3.4.2 *Implementation of the Shrinkage Model*

The definition of the general parameters (Fig. 3.3) is the same as before. The model parameters are shown in Fig. 3.7, the internal variables of the model are shown in Fig. 3.8. The capillary pressure (shrinkage stress) was applied as a pore-pressure (Fig. 3.9), regarding the poro-mechanical theory, using the *External Stress* subnode which allows the user to specify an additional stress contribution to the material which is not part of the constitutive relation. The external stress can be added to the total stress tensor, or act only as an extra load contribution. When the *Pore Pressure* option is selected, there is no contribution to the stress tensor and the only effect of the pressure is as a load.

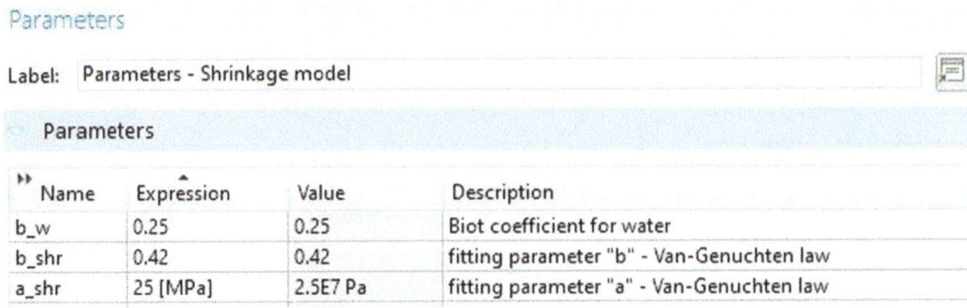

Name	Expression	Value	Description
b_w	0.25	0.25	Biot coefficient for water
b_shr	0.42	0.42	fitting parameter "b" - Van-Genuchten law
a_shr	25 [MPa]	2.5E7 Pa	fitting parameter "a" - Van-Genuchten law

Fig. 3.7 Shrinkage model parameters

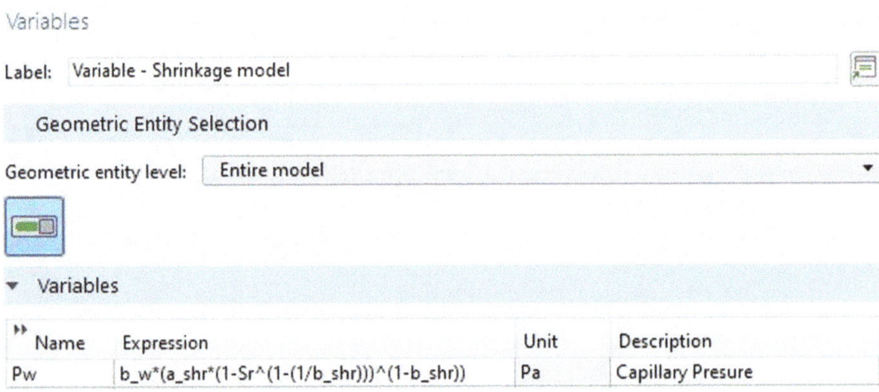

Name	Expression	Unit	Description
Pw	b_w*(a_shr*(1-Sr^(1-(1/b_shr)))^(1-b_shr))	Pa	Capillary Presure

Fig. 3.8 Internal variables of the shrinkage model

3.4.3 *Implementation of the Creep Model*

The definition of the general parameters (Fig. 3.3) is the same as before. First, it's presented the implementation of the Maxwell module (permanent creep) and after it's presented the implementation of the Kelvin module (reversible creep). Both strains were applied using the *External Strain* subnode which allows the user to provide inelastic strain contributions to the material on a variety of formats, including using external coded functions. For both modules (Maxwell and Kelvin), the external coded functions were implemented as ODEs to calculate the strains components in the main directions ($a_{ij} = a_{11}$, a_{22}) regarding the 2D application.

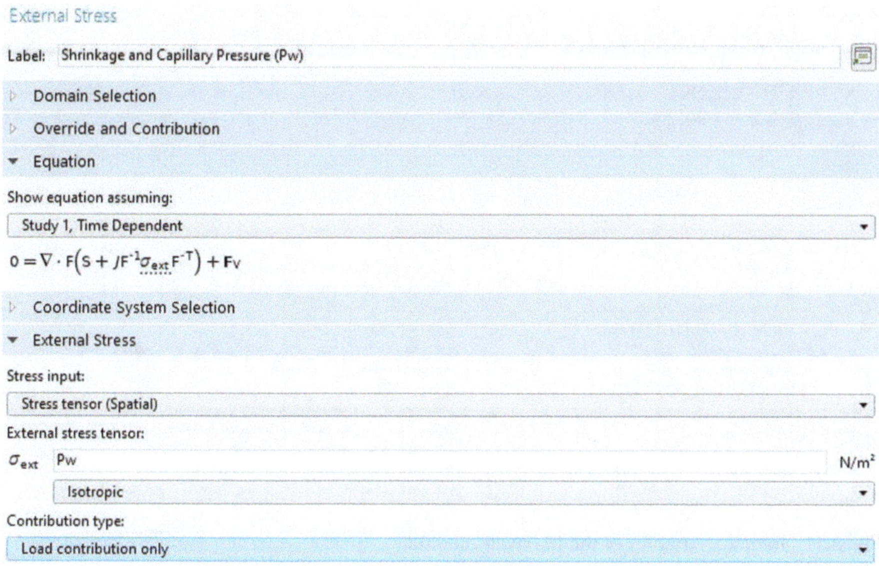

Fig. 3.9 Capillary pressure (shrinkage stress)

Maxwell Module (Permanent Creep)

The model parameters are shown in Fig. 3.10, the internal variables of the model are shown in Fig. 3.11. The internal variables for the function to consider the non-linearity dependence on loading levels are shown in Fig. 3.12. The coded function (ODE) to calculate the Maxwell strain is presented in Fig. 3.13 (discretization) and Fig. 3.14 (code), and the main components used as *External Strain* are shown in Fig. 3.15.

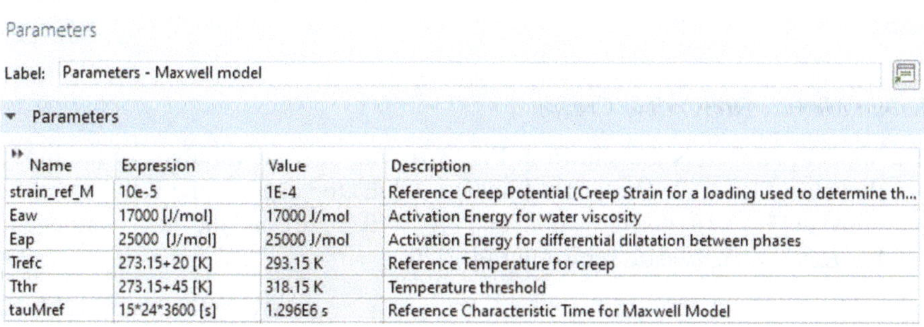

Name	Expression	Value	Description
strain_ref_M	10e-5	1E-4	Reference Creep Potential (Creep Strain for a loading used to determine th...
Eaw	17000 [J/mol]	17000 J/mol	Activation Energy for water viscosity
Eap	25000 [J/mol]	25000 J/mol	Activation Energy for differential dilatation between phases
Trefc	273.15+20 [K]	293.15 K	Reference Temperature for creep
Tthr	273.15+45 [K]	318.15 K	Temperature threshold
tauMref	15*24*3600 [s]	1.296E6 s	Reference Characteristic Time for Maxwell Model

Fig. 3.10 Maxwell module parameters

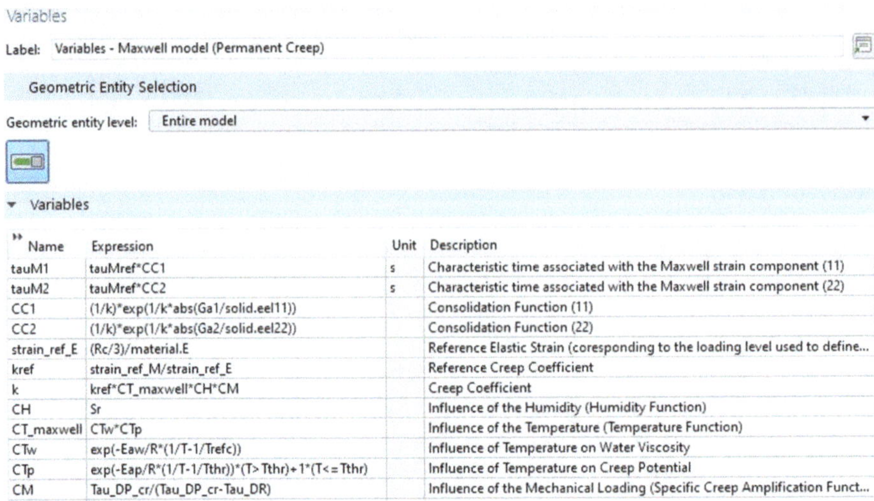

Name	Expression	Unit	Description
tauM1	tauMref*CC1	s	Characteristic time associated with the Maxwell strain component (11)
tauM2	tauMref*CC2	s	Characteristic time associated with the Maxwell strain component (22)
CC1	(1/k)*exp(1/k*abs(Ga1/solid.eel11))		Consolidation Function (11)
CC2	(1/k)*exp(1/k*abs(Ga2/solid.eel22))		Consolidation Function (22)
strain_ref_E	(Rc/3)/material.E		Reference Elastic Strain (coresponding to the loading level used to define...
kref	strain_ref_M/strain_ref_E		Reference Creep Coefficient
k	kref*CT_maxwell*CH*CM		Creep Coefficient
CH	Sr		Influence of the Humidity (Humidity Function)
CT_maxwell	CTw*CTp		Influence of the Temperature (Temperature Function)
CTw	exp(-Eaw/R*(1/T-1/Trefc))		Influence of Temperature on Water Viscosity
CTp	exp(-Eap/R*(1/T-1/Tthr))*(T>Tthr)+1*(T<=Tthr)		Influence of Temperature on Creep Potential
CM	Tau_DP_cr/(Tau_DP_cr-Tau_DR)		Influence of the Mechanical Loading (Specific Creep Amplification Funct...

Fig. 3.11 Internal variables of the Maxwell module

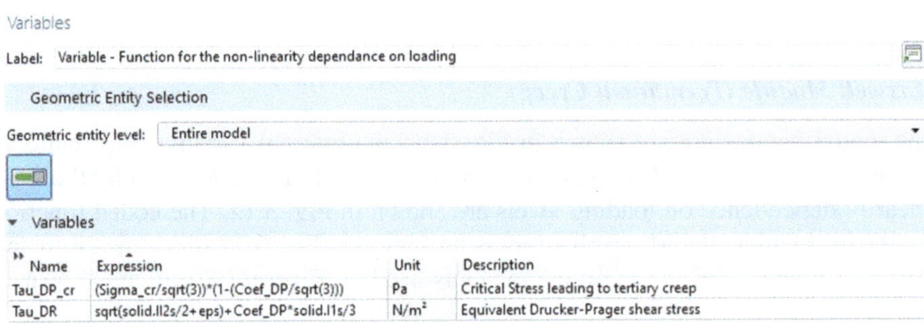

Name	Expression	Unit	Description
Tau_DP_cr	(Sigma_cr/sqrt(3))*(1-(Coef_DP/sqrt(3)))	Pa	Critical Stress leading to tertiary creep
Tau_DR	sqrt(solid.II2s/2+eps)+Coef_DP*solid.I1s/3	N/m²	Equivalent Drucker-Prager shear stress

Fig. 3.12 Internal variables of the function to consider the non-linearity dependance on loading level

Kelvin Module (Reversible Creep)

The model parameters are shown in Fig. 3.16, the internal variables of the model are shown in Fig. 3.17. The coded function (ODE) to calculate the Kelvin strain is presented in Fig. 3.18 (discretization) and Fig. 3.19 (code), and the main components used as *External Strain* are shown in Fig. 3.20.

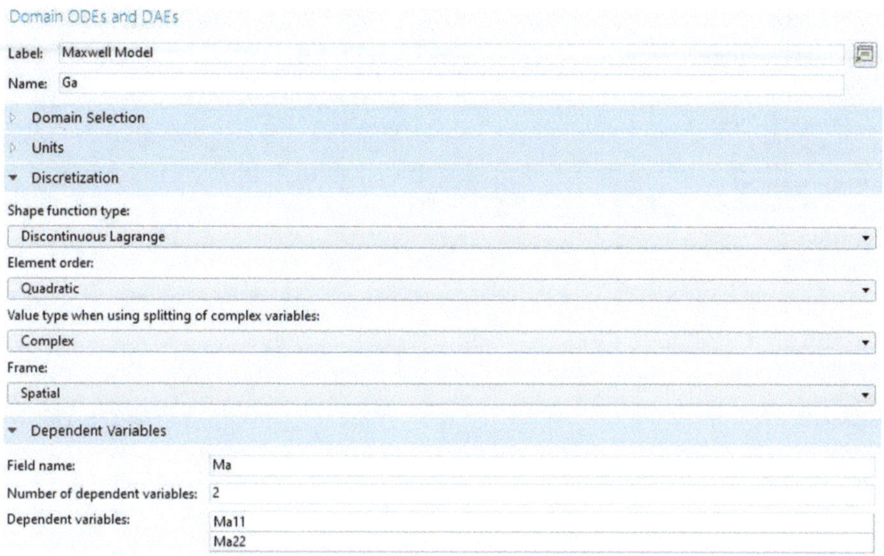

Fig. 3.13 Discretization of the Maxwell module applied as an ODE

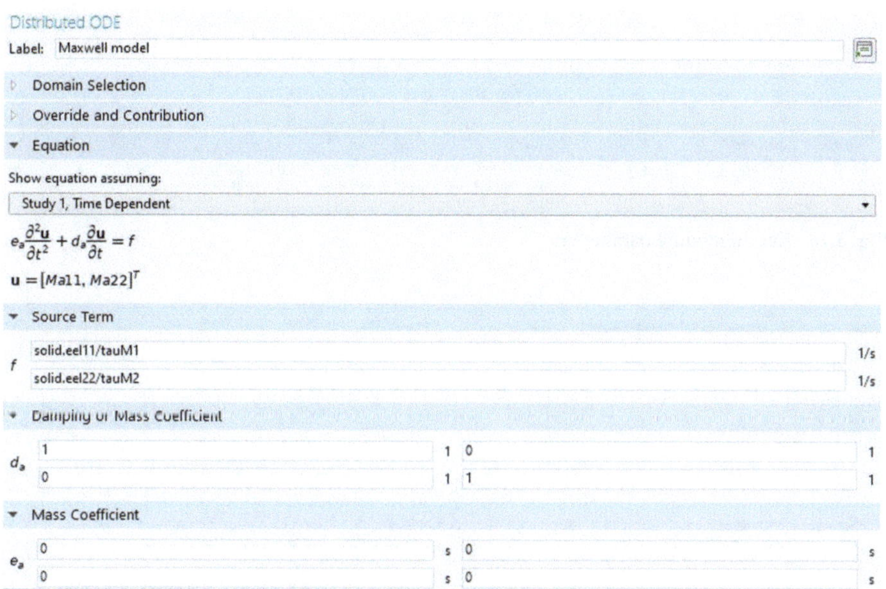

Fig. 3.14 Coded function of the Maxwell module applied as an ODE

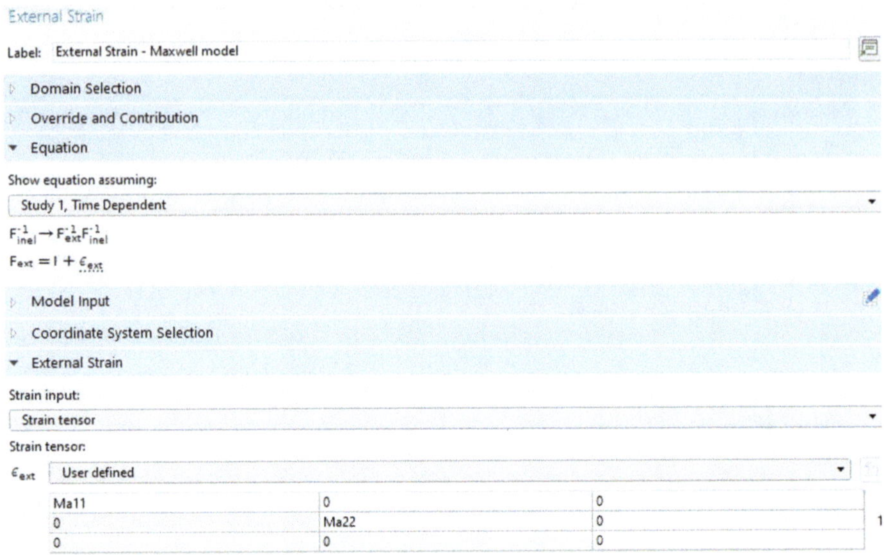

Fig. 3.15 Maxwell strains applied External Strain

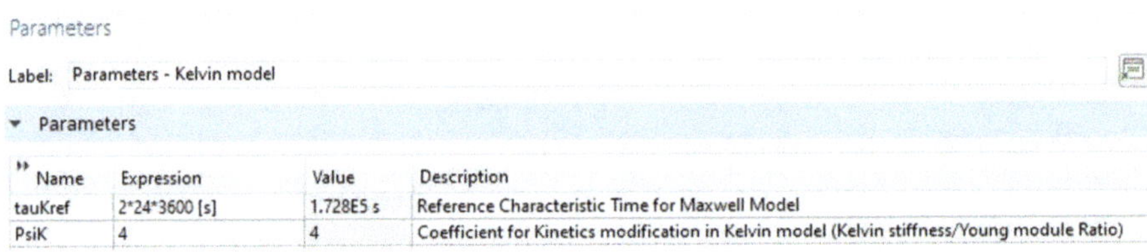

Fig. 3.16 Kelvin module parameters

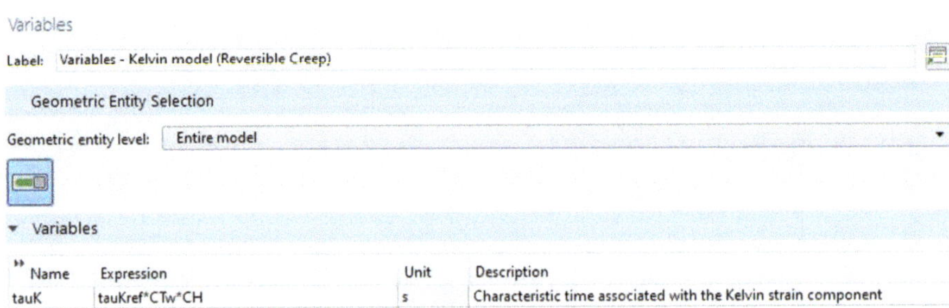

Fig. 3.17 Internal variables of the Kelvin module

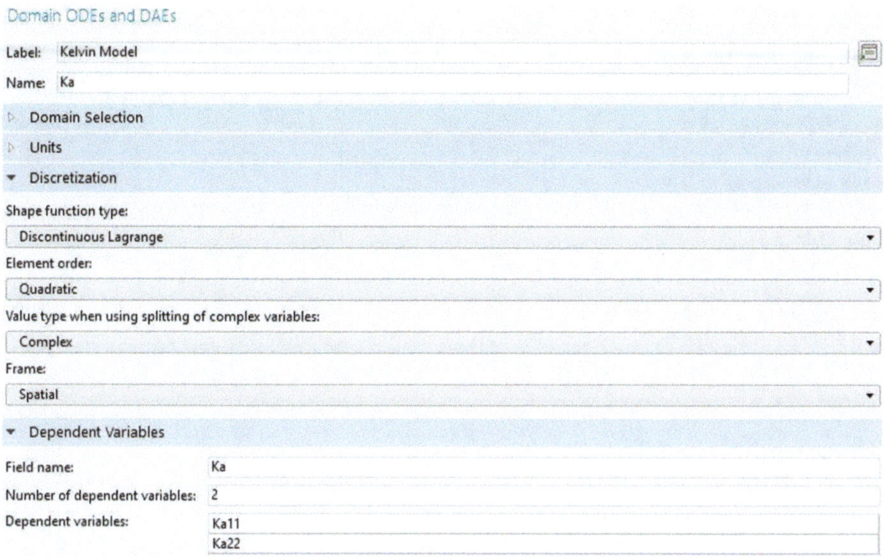

Fig. 3.18 Discretization of the Kelvin module applied as an ODE

Fig. 3.19 Coded function of the Kelvin module applied as an ODE

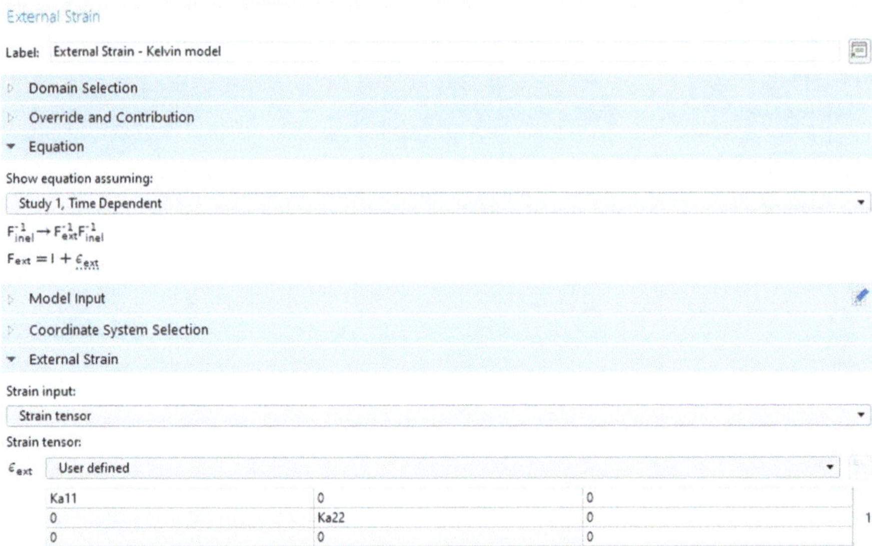

Fig. 3.20 Kelvin strains applied External Strain

References

Baroghel-Bouny V, Mainguy M, Lassabatere T, Coussy O (1999) Characterization and identification of equilibrium and transfer moisture properties for ordinary and high-performance cementitious materials. Cem Concr Res 29:1225–1238

Grimal E, Sellier A, Le Pape Y, Bourdarot E (2007) Creep, shrinkage, and anisotropic damage in alkali-aggregate reaction swelling mechanism—part I: a constitutive model. ACI Mater J 227–235

Grimal E, Morenon P, Sellier A, Multon S, Bourdarot E (2017) AAR and DEF structural effects modelling. In: Swelling concrete in dams and hydraulic structures: DSC 2017, 1st edn. ISTE Ltd., Wiley

Larive C (1997) Apports combinés de l'expérimentation et de la modélisation à la compréhension de l'alcali-réaction et de ses effets mécaniques. Ph.D. thesis, Ecole Nationale des Ponts et Chaussées, Paris, p 335

Morenon P (2017) Modélisation des reactions de gonflement interne des bétons avec prise en compte des couplages poro-mécaniques et chimiques. Ph.D. thesis, Université Toulouse 3 Paul Sabatier, France, p 272

Multon S, Toutlemonde F (2006) Effect of applied stresses on alkali-silica reaction induced expansions. Cem Concr Res 36:912–920

Poyet S, Sellier A, Capra B, Thèvenin-Foray G, Torrenti J-M, Tournier-Cognon H, Bourdarot E (2006) Influence of water on alkali-silica reaction: experimental study and numerical simulations. J Mater Civ Eng 18:588–596

Sellier A, Casaux-Ginestet G, Buffo-Lacarrière L, Bourbon X (2013) Orthotropic damage coupled with localized crack reclosure processing—part I: constitutive laws. Eng Fract Mech 971:148–167

Sellier A, Multon S, Buffo-Lacarrière L, Vidal T, Bourbon X, Guillaume C (2016) Concrete creep modelling for structural applications: non-linearity, multi-axiality, hydration, temperature and drying effects. Cem Concr Res 79:301–315

Ulm F-J, Coussy O, Larive C, Kefei L (2000) Thermo-chemo-mechanics of ASR expansion in concrete structures. J Eng Mech 126:233–242

Chapter 4
Results and Discussion

In this chapter is presented the main simulations and the respective results of the models separately. The implementation in COMSOL was done by two different types of calibration: theoretical and experimental.

First, is presented the theoretical evolution of the advancement of the reaction in a parametric study, comparing with the results shown by Morenon (2017). The second part is dedicated to present the results of the creep model. First is presented the theoretical behavior of a concrete specimen subjected to a loading, followed by unloading, compared to the theoretical behavior shown by Mehta and Monteiro (2014)—see Fig. 2.26. Next, the creep model was compared to the numerical benchmark shown by Sellier et al. (2016), which is focused in the comparison with experimental data found in the literature. An advantage of this procedure is the fact that the parameters were already calibrated by the developer of the model.

For the sake of simplicity, the chemical model was evaluated through simulations in a 0D framework, and for the creep model, the simulations were done in a 2D framework.

4.1 Results of the Chemical Model

In order to evaluate the implementation of the chemical model in COMSOL Multiphysics®, it was analyzed the evolution of the advancement of the reaction (A^{asr}), the main vector in the chemical model to produce a volume of gel (Morenon 2017). It was done a parametric study regarding to the influence of the three main parameters which influences on the advancement of the reaction: (i) characteristic time (τ_{ref}^{asr}); (ii) temperature (T); (iii) saturation degree (Sr). The main material parameters are available in Table 4.1.

This first approach was modeled in a 0D framework, without explicitly modeling a material such as concrete. Actually, the material is substituted by a point, and the

F. A. N. Silva et al., *Advanced Chemical and Creep Modeling for Alkali-Aggregate Reaction in Concrete*, Building Pathology and Rehabilitation 28,
https://doi.org/10.1007/978-3-031-53980-0_4

Table 4.1 Loading scenarios for the multiaxial creep tests by Kim et al. (2005)

Concrete type	Uniaxial compression		Biaxial compression		
	U1 (MPa)	U2 (MPa)	B1 (MPa)	B2 (MPa)	B3 (MPa)
C1	$\sigma_I = 4.9$	$\sigma_I = 9.8$	$\sigma_I = 4.9$ $\sigma_{II} = 0.98$	$\sigma_I = 4.9$ $\sigma_{II} = 1.96$	$\sigma_I = 9.8$ $\sigma_{II} = 1.96$
C2	$\sigma_I = 7.35$	$\sigma_I = 9.8$	$\sigma_I = 7.35$ $\sigma_{II} = 1.47$	$\sigma_I = 7.35$ $\sigma_{II} = 2.94$	$\sigma_I = 9.8$ $\sigma_{II} = 2.94$
C3	$\sigma_I = 9.8$	$\sigma_I = 12.25$	$\sigma_I = 9.8$ $\sigma_{II} = 1.96$	$\sigma_I = 9.8$ $\sigma_{II} = 3.92$	$\sigma_I = 12.25$ $\sigma_{II} = 3.92$

chemical response is observed regardless of the material behavior. It was done in a way to analyze and validate the chemical model only.

In Fig. 4.1, it's presented the parametric study on the characteristic time (τ_{ref}^{asr}), which this material parameter was set in 50, 100, 150 and 200 days. The saturation degree was set in 100% and the temperature was set in 38 °C. The evolution of A^{asr} is in very good agreement with the results shown by Morenon (2017).

In Fig. 4.2, it's presented the parametric study on the temperature (T), which this environmental parameter was set in 10, 20, 30 and 40 °C. The saturation degree was set in 100% and the characteristic time was set in 50 days. The evolution of A^{asr} is in very good agreement with the results shown by Morenon (2017).

In Fig. 4.3, it's presented the parametric study on the saturation degree (Sr), which this environmental parameter was set in 25, 50, 75 and 100%. The temperature was

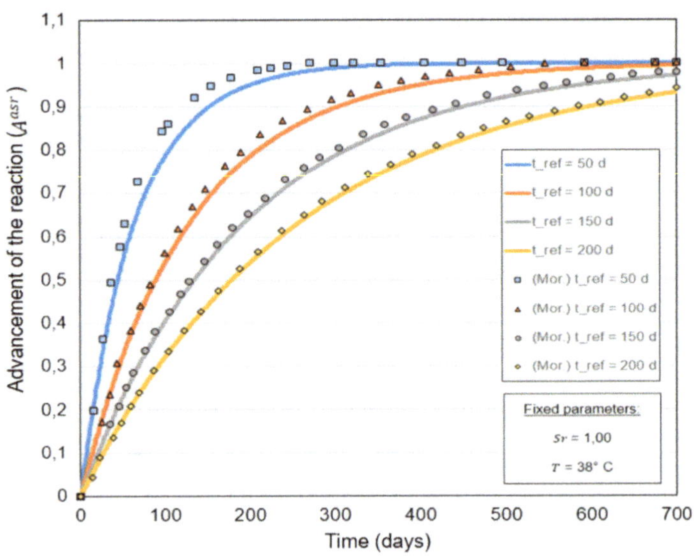

Fig. 4.1 Effects of the characteristic time over the advancement of the reaction—comparison with the results shown by Morenon (2017)

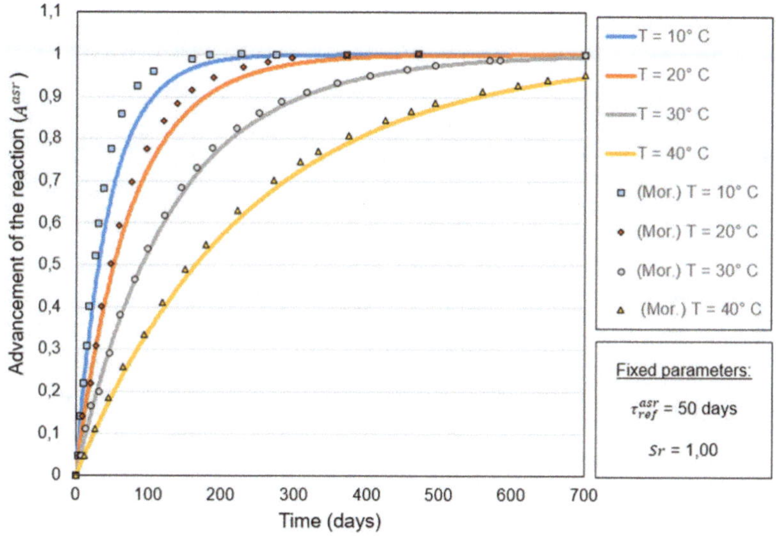

Fig. 4.2 Effects of the temperature over the advancement of the reaction—comparison with the results shown by Morenon (2017)

set in 38 °C and the characteristic time was set in 50 days. The evolution of A^{asr} is slower than shown by Morenon (2017). This discrepancy can be due to an alteration of the parameters presented by Morenon, because the curves for the same parameters ($Sr = 100\%$, $T = 38\,°\text{C}$, $\tau_{ref}^{asr} = 50\,\text{days}$) don't match, as seen in Fig. 4.4.

4.2 Results of the Creep Model

In order to evaluate the implementation of the creep model in COMSOL Multiphysics®, it was necessary to simulate the creep tests performed by Kim et al. (2005) and Ladaoui et al. (2013). Sellier et al. (2013) calibrated the main material parameters and compared the results of their model with the experimental data, and the results presented were in very good agreement.

4.2.1 Multiaxial Creep Test

The tests performed by Kim et al. (2005) consist of multiaxial creep tests of 27 cubic concrete specimens (200 mm × 200 mm × 200 mm) with different strengths (26, 44 and 54 MPa), and the creep strains were measured in three principal directions. The tests were divided in uniaxial compression, biaxial compression and triaxial

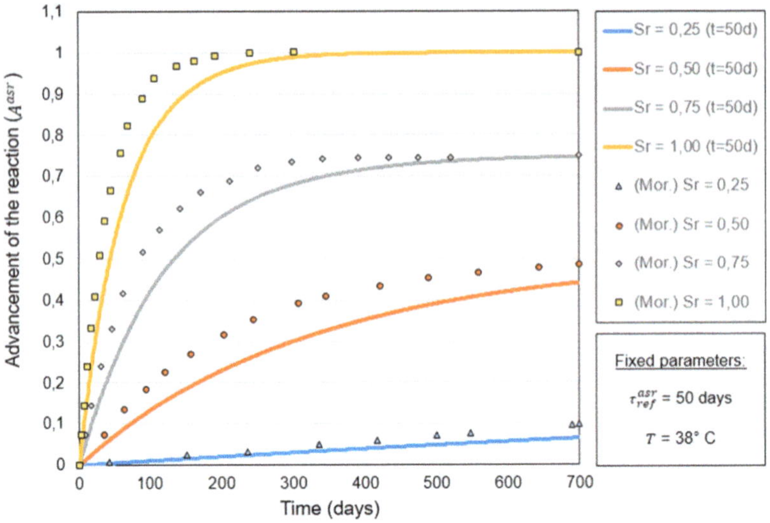

Fig. 4.3 Effects of the saturation degree over the advancement of the reaction—comparison with the results shown by Morenon (2017)

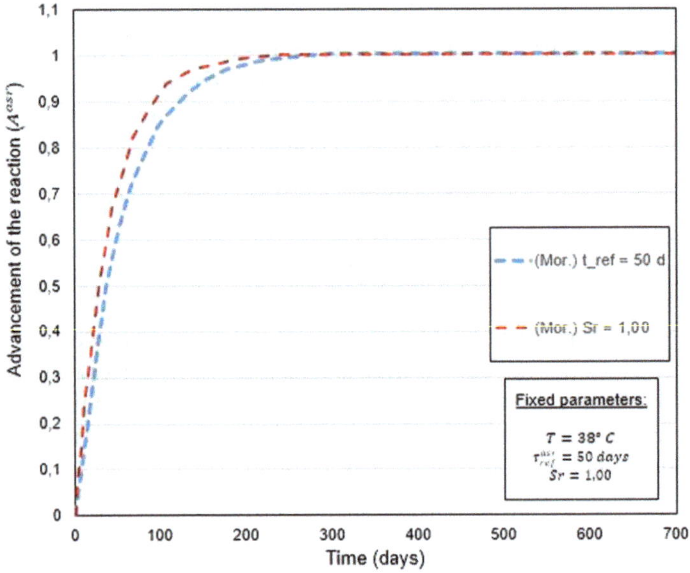

Fig. 4.4 Difference of the evolution of the advancement of the reaction for the same main parameters, shown by Morenon (2017)

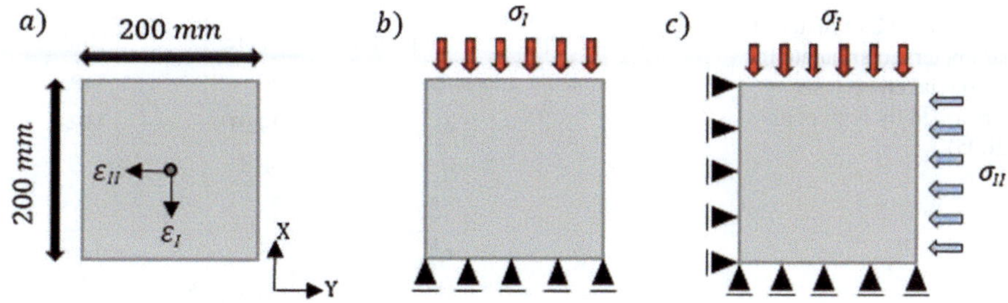

Fig. 4.5 **a** Geometry of the problem; **b** boundary conditions for uniaxial compression; **c** boundary conditions for biaxial compression

compression. For the sake of simplicity, the numerical simulation was performed only for uniaxial and biaxial compression, as the simulation was done in a 2D framework.

The geometry and boundary conditions of the simulation are presented in Fig. 4.5, for both uniaxial (Fig. 4.5b) and biaxial compression (Fig. 4.5c). The test was performed for three different types of concrete (C1, C2 and C3) which had different material parameters. The loading scenarios varied for each type of concrete and also for uniaxial (U1 and U2) and biaxial compression (B1, B2 and B3). Next, on Table 4.1, it's presented the loading configuration for each scenario and type of concrete.

Next, on Table 4.2, there is a resume of the calibrated parameters used to simulate the creep tests performed by Kim et al. (2005).

The results for the uniaxial compression for concrete C1, C2 and C3 are presented in Figs. 4.6, 4.7 and 4.8, respectively. The maximum discrepancy between the simulation and experimental results is approximately 10%, which is considered satisfactory for numerical simulation of experimental benchmark.

The results for the biaxial compression for concrete C1, C2 and C3 are presented in Figs. 4.9, 4.10 and 4.11, respectively. The maximum discrepancy was approximately 13%, very close to the one found in uniaxial compression comparison.

The discrepancy observed on the results of the strain ε_I—parallel to the load σ_I—was higher than observed on the strain ε_{II}—parallel to the load σ_{II}. It can be inferred that higher loads induces to higher discrepancy, as σ_I was 2.5 and 5 times higher than σ_{II}. Also it can be seen that the discrepancy is higher for longer periods of time, as the creep strain increases with time.

4.2.2 *Uniaxial Creep Test in Different Temperature Conditions*

The tests performed by Ladaoui et al. (2013) consist in a uniaxial creep test in different temperature conditions of 24 cylindrical specimens (220 mm × 110 mm) of high performance concrete (HPC), with 86 MPa of compressive strength and 45 MPa

Table 4.2 Main material parameters to simulate the multiaxial creep tests performed by Kim et al. (2005)

Concrete type	Parameter	Value	Unit
C1	Rc	26	MPa
	E	24.010	MPa
	υ	0.18	–
	τ_{ref}^{K}	2	days
	τ_{ref}^{M}	15	days
	ε_{ref}^{M}	$10 \times e^{-5}$	–
	ψ^{K}	4	–
	ε_{ref}^{E}	$211 \times e^{-6}$	–
C2	Rc	44.1	MPa
	E	29.841	MPa
	υ	0.18	–
	τ_{ref}^{K}	2	days
	τ_{ref}^{M}	15	days
	ε_{ref}^{M}	$9 \times e^{-5}$	–
	ψ^{K}	4	–
	ε_{ref}^{E}	$246 \times e^{-6}$	–
C3	Rc	54.3	MPa
	E	34.006	MPa
	υ	0.18	–
	τ_{ref}^{K}	3	days
	τ_{ref}^{M}	20	days
	ε_{ref}^{M}	$7.5 \times e^{-5}$	–
	ψ^{K}	4	–
	ε_{ref}^{E}	$288 \times e^{-6}$	–

of Young's modulus. The creep strains were measured in the principal direction (direction of the loading). After one year of curing, an uniaxial compression stress of 26 MPa was applied for specimens under 20, 50 and 80 °C. For the sake of simplicity, the numerical simulation was performed only for uniaxial and biaxial compression, because it was done in a 2D-axisymetric framework. The geometry and boundary conditions of the simulation are presented in Fig. 4.12.

Next, on Table 4.3, there is a resume of the calibrated parameters used to simulate the creep tests performed by Ladaoui et al. (2013).

The influence of the temperature in the creep tests can be visualized through the results of the simulation for the temperatures of 20, 50 and 80 °C. The results are presented in Fig. 4.13. The experimental results for the temperature of 80 °C after

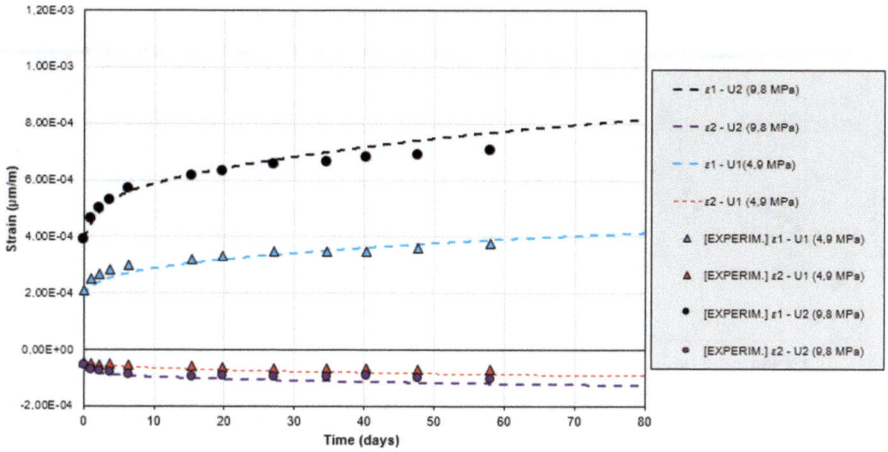

Fig. 4.6 Evolution of the total strain for concrete type C1 under uniaxial compression

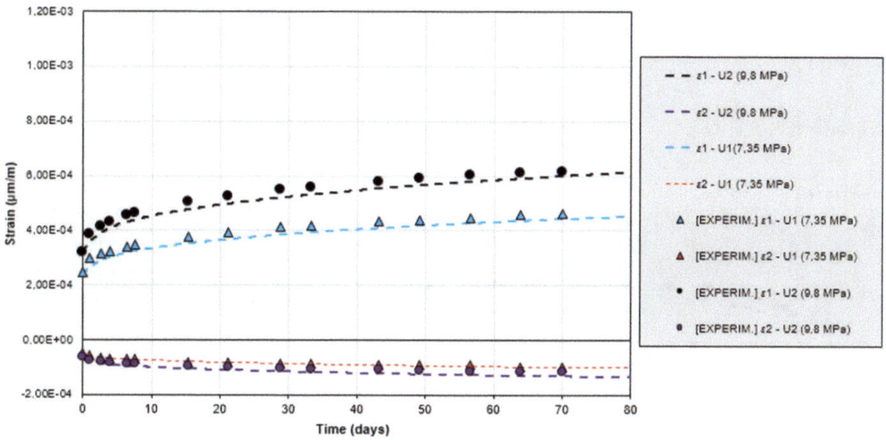

Fig. 4.7 Evolution of the total strain for concrete type C2 under uniaxial compression

56 days were not given due to a malfunction in the strain measurer, because the high temperature condition led to a failure on the electric parts of the measurer.

The curves for 20 and 50 °C present higher discrepancy for longer periods of time, as expected regarding the results for multiaxial creep test. The maximum discrepancy was approximately 6% which can be considered in a very good agreement with the experimental benchmark.

In the other hand, the curves for 80 °C already presents discrepancy of approximately 5% already at the early days (56 days). If the experimental results were available, it would be possible to check if the maximum discrepancy overtime would be much higher than the ones calculated for the curves of the other temperature conditions.

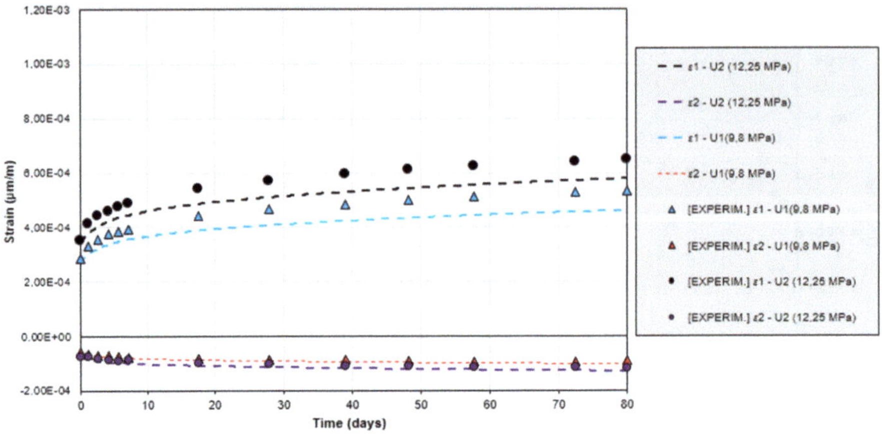

Fig. 4.8 Evolution of the total strain for concrete type C3 under uniaxial compression

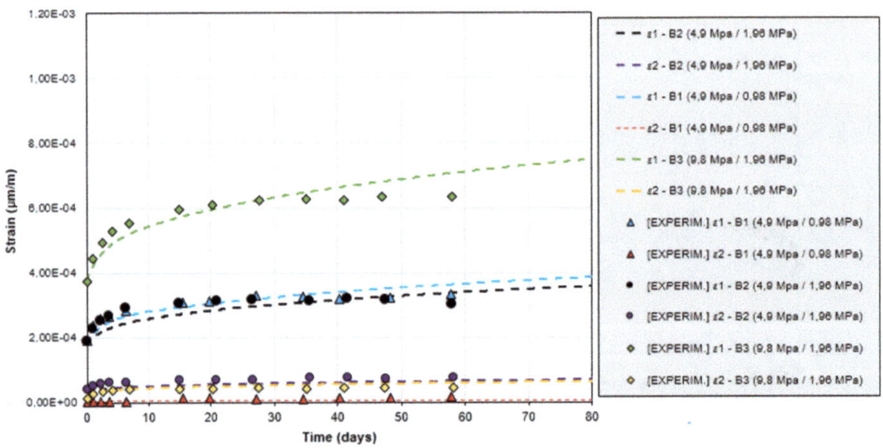

Fig. 4.9 Evolution of the total strain for concrete type C1 under biaxial compression

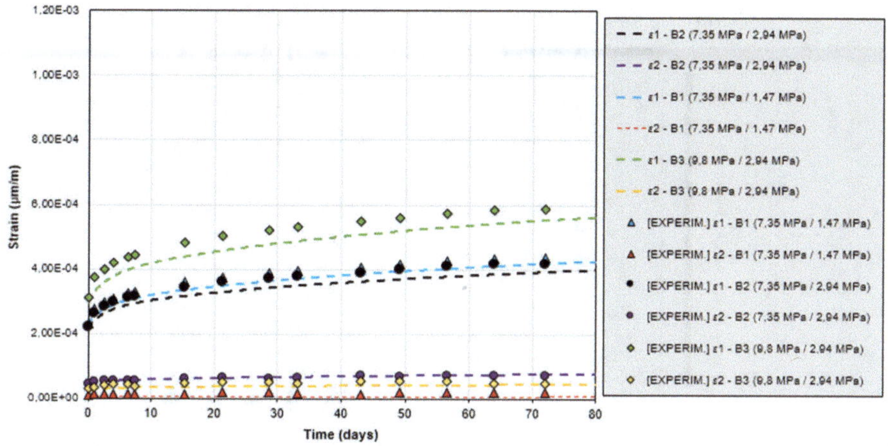

Fig. 4.10 Evolution of the total strain for concrete type C2 under biaxial compression

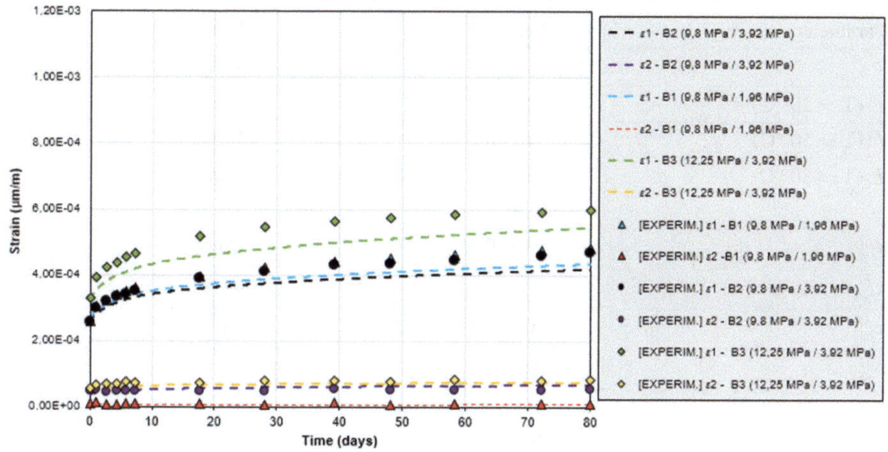

Fig. 4.11 Evolution of the total strain for concrete type C3 under biaxial compression

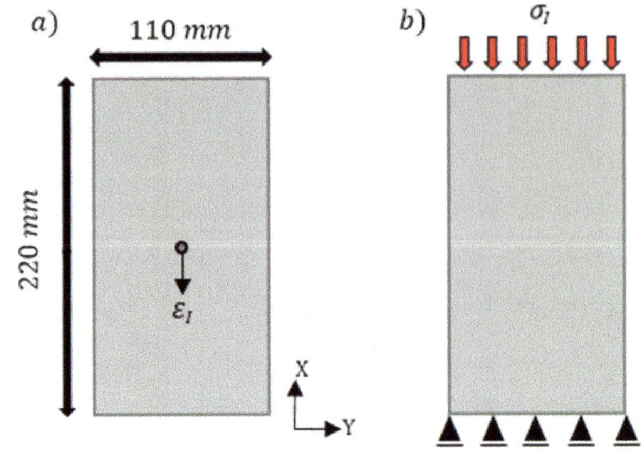

Fig. 4.12 **a** Geometry of the problem; **b** boundary conditions for uniaxial compression

Table 4.3 Main material parameters to simulate the creep tests performed by Ladaoui et al. (2013)

Parameter	Value	Unit
Rc	86	MPa
E (T = 20 °C)	34.800	MPa
E (T = 50 °C)	41.600	MPa
E (T = 80 °C)	45.450	MPa
υ	0.28	–
E_p^a	23.700	J/mol
τ_{ref}^K	5	days
τ_{ref}^M	10	days
ε_{ref}^M	5.5 × e^{-5}	–
ψ^K	5	–
ε_{ref}^E	590 × e^{-6}	–

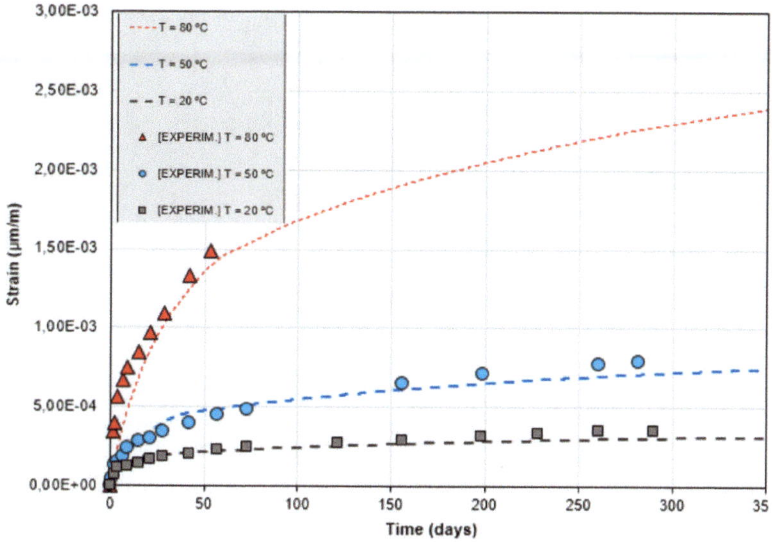

Fig. 4.13 Evolution of the total strain for the different temperatures in the creep test performed by Ladaoui et al. (2013)

References

Kim SY, Kim JK, Kim YY, Kwon SH (2005) Experimental studies on creep of sealed concrete under multiaxial stresses. Mag Concr Res 57:623–634

Ladaoui W, Vidal T, Sellier A, Bourbon X (2013) Analysis of interactions between damage and basic creep of HPC and HPFRC heated between 20 and 80 °C. Mater Struct 46:13–23

Mehta PK, Monteiro PJM (2014) Concreto: estrutura, propriedades e materiais, 2ª edn. In: Hasparyk NP, Helene P, Paulon VA (Reviewers) Translate of the 4th edition in English with Concrete: microstructure, properties and materials. IBRACON, São Paulo

Morenon P (2017) Modélisation des reactions de gonflement interne des bétons avec prise en compte des couplages poro-mécaniques et chimiques. Ph.D. thesis, Université Toulouse 3 Paul Sabatier, France, p 272

Sellier A, Casaux-Ginestet G, Buffo-Lacarrière L, Bourbon X (2013) Orthotropic damage coupled with localized crack reclosure processing part I: constitutive laws. Eng Fract Mech 971:148–167

Sellier A, Multon S, Buffo-Lacarrière L, Vidal T, Bourbon X, Guillaume C (2016) Concrete creep modelling for structural applications: non-linearity, multi-axiality, hydration, temperature and drying effects. Cem Concr Res 79:301–315

Chapter 5
Conclusions and Recommendations

In this chapter is discussed the main conclusions about the results presented on the previous chapter, and also about the limitations of the work. At the end, some suggestions and perspectives are presented in order to serve as a guideline for the next works in this same subject.

Analyzing the results, both chemical and creep models seem to work properly well, with low error values for numerical simulations. The method to compare the simulation results with the experimental benchmark—proposed by the developers of the models—guarantees that the implementation of such models in COMSOL Multiphysics® was a success.

Although such models require a huge number of material parameters that must be calibrated by the user. The values of those parameters can be found in the papers published by the developers of the models, but to implement in a real concrete structure assessment, there must be done a careful investigation over the structure. If those models are used still in the design phase of such structures, a parametric study is necessary to verify the variation in the mechanical response due to the variation of such material parameters.

First, as the reader was capable to observe, the chemical and creep model were not coupled. The attempts to do this result in numerical instability regarding time. The author was not capable to verify the reason behind it, neither identify which condition must be satisfied in order to do this.

Secondly, the pressure of the gel was not calculated, because to do so is necessary the creep strains and also the diffuse cracking (micro-plastic cracks) in order to generate this intrapore pressure.

Thirdly, the shrinkage model was not resulting in any significant strain, therefore it was not even compared to any experimental benchmark. Probably is due to the increment on the effective stress, which formulation in COMSOL Multiphysics® seems to be different from the formulation proposed by Sellier et al. (2016). More tests must be done to look carefully how to reproduce shrinkage in concrete materials.

F. A. N. Silva et al., *Advanced Chemical and Creep Modeling for Alkali-Aggregate Reaction in Concrete*, Building Pathology and Rehabilitation 28,
https://doi.org/10.1007/978-3-031-53980-0_5

Finally, to fully reproduce a concrete structure subjected to AAR, is necessary to implement a coupled chemo-creep-damage model. The author tried to use the isotropic damage model which was already developed in COMSOL Multiphysics® (*Damage* subnode for elastic materials). The creep model somehow doesn't converge with the damage model, mostly because numerical instability over time due to plastic strains interacting with creep strains. A lot of effort was put on the resolution of this problem, but no significant results were achieved.

Therefore, the next works in this subject must have as the principal aim the coupling of such models, comparing with numerical and experimental benchmark, firstly in a 2D approach for the sake of simplicity, and once the numerical stability is reached, the models must be tested in a 3D framework. The first approach on more complex simulations regarding AAR-affected concrete structures must verify the experimental and numerical benchmark, found in the works of Morenon (2017) and Sellier et al. (2016).

The author of this work is currently working on the execution of subnodes for COMSOL, using the Model Builder tool in order to pre-set the equations and set of parameters, making them available for all users in an easy way, regarding all the careful measures which are necessary to use such models.

References

Morenon P (2017) Modélisation des reactions de gonflement interne des bétons avec prise en compte des couplages poro-mécaniques et chimiques. Ph.D. thesis, Université Toulouse 3 Paul Sabatier, France, p 272

Sellier A, Multon S, Buffo-Lacarrière L, Vidal T, Bourbon X, Guillaume C (2016) Concrete creep modelling for structural applications: non-linearity, multi-axiality, hydration, temperature and drying effects. Cem Concr Res 79:301–315

MIX
Papier aus verantwortungsvollen Quellen
Paper from responsible sources
FSC® C105338

If you have any concerns about our products,
you can contact us on
ProductSafety@springernature.com

In case Publisher is established outside the EU,
the EU authorized representative is:
Springer Nature Customer Service Center GmbH
Europaplatz 3, 69115 Heidelberg, Germany

Printed by Libri Plureos GmbH
in Hamburg, Germany